Theodoros Katerinakis

The Social Construction of Knowledge in Mission-Critical Environments

Lessons from the Flight Deck

Springer

Theodoros Katerinakis
Drexel On-Line Council
Drexel University
Philadelphia, PA, USA

ISSN 2197-5698 ISSN 2197-5701 (electronic)
Innovation, Technology, and Knowledge Management
ISBN 978-3-030-08156-0 ISBN 978-3-319-91014-7 (eBook)
https://doi.org/10.1007/978-3-319-91014-7

Printed on acid-free paper

This Springer imprint is published by the registered company Springer International Publishing AG part of Springer Nature.
The registered company address is: Gewerbestrasse 11, 6330 Cham, Switzerland

Innovation, Technology, and Knowledge Management

Series Editor

Elias G. Carayannis
George Washington University
Washington, DC, USA

More information about this series at http://www.springer.com/series/8124

Series Foreword

The Springer book series *Innovation, Technology, and Knowledge Management* was launched in March 2008 as a forum and intellectual, scholarly "podium" for global/local, transdisciplinary, transsectoral, public–private, and leading/"bleeding"-edge ideas, theories, and perspectives on these topics.

The book series is accompanied by the Springer *Journal of the Knowledge Economy*, which was launched in 2009 with the same editorial leadership.

The series showcases provocative views that diverge from the current "conventional wisdom," that are properly grounded in theory and practice, and that consider the concepts of *robust competitiveness*,[1] *sustainable entrepreneurship*,[2] and *democratic capitalism*[3] central to its philosophy and objectives. More specifically, the aim of this series is to highlight emerging research and practice at the dynamic intersection of these fields, where individuals, organizations, industries, regions, and nations are harnessing creativity and invention to achieve and sustain growth.

---

[1] We define *sustainable entrepreneurship* as the creation of viable, profitable, and scalable firms. Such firms engender the formation of self-replicating and mutually enhancing innovation networks and knowledge clusters (innovation ecosystems), leading toward robust competitiveness (E.G. Carayannis, *International Journal of Innovation and Regional Development* 1(3), 235–254, 2009).

[2] We understand *robust competitiveness* to be a state of economic being and becoming that avails systematic and defensible "unfair advantages" to the entities that are part of the economy. Such competitiveness is built on mutually complementary and reinforcing low-, medium-, and hightechnology and public and private sector entities (government agencies, private firms, universities, and nongovernmental organizations) (E.G. Carayannis, *International Journal of Innovation and Regional Development* 1(3), 235–254, 2009).

[3] The concepts of *robust competitiveness* and *sustainable entrepreneurship* are pillars of a regime that we call *democratic capitalism* (as opposed to "popular or casino capitalism"), in which real opportunities for education and economic prosperity are available to all, especially—but not only—younger people. These are the direct derivative of a collection of top-down policies as well as bottom-up initiatives (including strong research and development policies and funding, but going beyond these to include the development of innovation networks and knowledge clusters across regions and sectors) (E.G. Carayannis and A. Kaloudis, *Japan Economic Currents*, pp. 6–10, January 2009).

Books that are part of the series explore the impact of innovation at the "macro" (economies, markets), "meso" (industries, firms), and "micro" levels (teams, individuals), drawing from such related disciplines as finance, organizational psychology, research and development, science policy, information systems, and strategy, with the underlying theme that for innovation to be useful, it must involve the sharing and application of knowledge.

Some of the key anchoring concepts of the series are outlined in the figure below and the definitions that follow (all definitions are from Carayannis and Campbell (2009)).

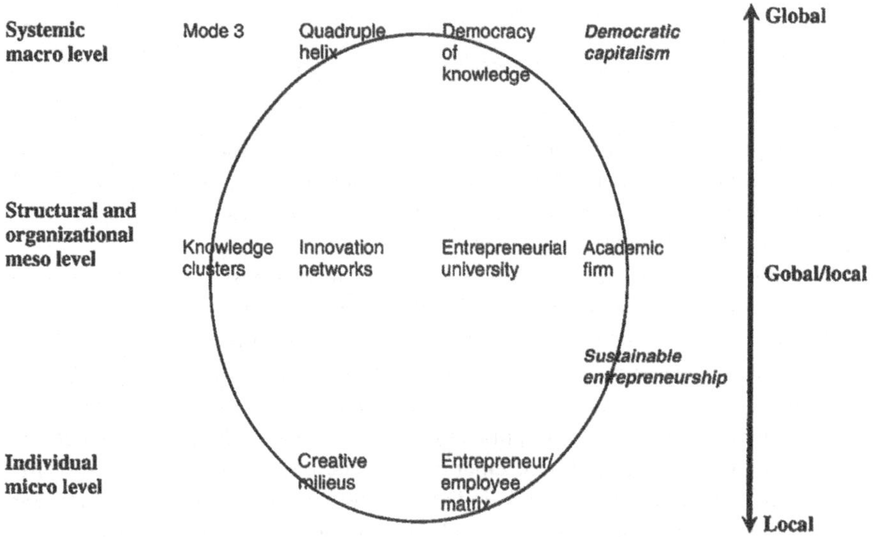

Conceptual profile of the series *Innovation, Technology, and Knowledge Management*:

- The "Mode 3" Systems Approach for Knowledge Creation, Diffusion, and Use: "Mode 3" is a multilateral, multinodal, multimodal, and multilevel systems approach to the conceptualization, design, and management of real and virtual, "knowledge-stock" and "knowledge-flow," modalities that catalyze, accelerate, and support the creation, diffusion, sharing, absorption, and use of cospecialized knowledge assets. "Mode 3" is based on a system-theoretic perspective of socio-economic, political, technological, and cultural trends and conditions that shape the coevolution of knowledge with the "knowledge-based and knowledge-driven, global/local economy and society."
- Quadruple Helix: Quadruple helix, in this context, means to add to the triple helix of government, university, and industry a "fourth helix" that we identify as the "media-based and culture-based public." This fourth helix associates with "media," "creative industries," "culture," "values," "lifestyles," "art," and perhaps also the notion of the "creative class."

- Innovation Networks: Innovation networks are real and virtual infrastructures and infratechnologies that serve to nurture creativity, trigger invention, and catalyze innovation in a public and/or private domain context (for instance, government–university–industry public–private research and technology development coopetitive partnerships).
- Knowledge Clusters: Knowledge clusters are agglomerations of cospecialized, mutually complementary, and reinforcing knowledge assets in the form of "knowledge stocks" and "knowledge flows" that exhibit self-organizing, learning-driven, dynamically adaptive competences and trends in the context of an open systems perspective.
- Twenty-First-Century Innovation Ecosystem: A twenty-first-century innovation ecosystem is a multilevel, multimodal, multinodal, and multiagent system of systems. The constituent systems consist of innovation metanetworks (networks of innovation networks and knowledge clusters) and knowledge metaclusters (clusters of innovation networks and knowledge clusters) as building blocks and organized in a self-referential or chaotic fractal knowledge and innovation architecture (Carayannis 2001), which in turn constitute agglomerations of human, social, intellectual, and financial capital stocks and flows as well as cultural and technological artifacts and modalities, continually coevolving, cospecializing, and cooperating. These innovation networks and knowledge clusters also form, reform, and dissolve within diverse institutional, political, technological, and socioeconomic domains, including government, university, industry, and nongovernmental organizations and involving information and communication technologies, biotechnologies, advanced materials, nanotechnologies, and next-generation energy technologies.

Who is this book series published for? The book series addresses a diversity of audiences in different settings:

1. *Academic communities*: Academic communities worldwide represent a core group of readers. This follows from the theoretical/conceptual interest of the book series to influence academic discourses in the fields of knowledge, also carried by the claim of a certain saturation of academia with the current concepts and the postulate of a window of opportunity for new or at least additional concepts. Thus, it represents a key challenge for the series to exercise a certain impact on discourses in academia. In principle, all academic communities that are interested in knowledge (knowledge and innovation) could be tackled by the book series. The interdisciplinary (transdisciplinary) nature of the book series underscores that the scope of the book series is not limited a priori to a specific basket of disciplines. From a radical viewpoint, one could create the hypothesis that there is no discipline where knowledge is of no importance.
2. *Decision-makers—private/academic entrepreneurs and public (governmental, subgovernmental) actors*: Two different groups of decision-makers are being addressed simultaneously: (1) private entrepreneurs (firms, commercial firms, academic firms) and academic entrepreneurs (universities), interested

in optimizing knowledge management and in developing heterogeneously composed knowledge-based research networks, and (2) public (governmental, subgovernmental) actors that are interested in optimizing and further developing their policies and policy strategies that target knowledge and innovation. One purpose of public *knowledge and innovation policy* is to enhance the performance and competitiveness of advanced economies.

3. *Decision-makers in general*: Decision-makers are systematically being supplied with crucial information, for how to optimize knowledge-referring and knowledge-enhancing decision-making. The nature of this "crucial information" is conceptual as well as empirical (case study-based). Empirical information highlights practical examples and points toward practical solutions (perhaps remedies); conceptual information offers the advantage of further-driving and further-carrying tools of understanding. Different groups of addressed decision-makers could be decision-makers in private firms and multinational corporations, responsible for the knowledge portfolio of companies; knowledge and knowledge management consultants; globalization experts, focusing on the internationalization of research and development, science and technology, and innovation; experts in university/business research networks; and political scientists, economists, and business professionals.
4. *Interested global readership*: Finally, the Springer book series addresses a whole global readership, composed of members who are generally interested in knowledge and innovation. The global readership could partially coincide with the communities as described above ("academic communities," "decision-makers"), but could also refer to other constituencies and groups.

Washington, DC, USA Elias G. Carayannis

References

Carayannis, E. G. (2001). *Strategic management of technological learning*. Boca Raton: CRC Press.

Carayannis, E. G., & Campbell, D. F. J. (2009). "Mode 3" and "Quadruple Helix": toward a 21st century fractal innovation ecosystem. *International Journal of Technology Management*, *46*(3–4), 201–234.

To Maria and Antonios-Emmanuel

Endorsements

In this path-breaking work, Theodore Katerinakis brings the study of human communication to the airplane cockpit as a knowledge environment. Toward that end, drawing on his own experience with the Air Force and aviation authorities and interviews with flight controllers and scores of pilots, Katerinakis both builds on and moves beyond human factors research and ecological psychology. In doing so, he demonstrates how important conversation dynamics are to safe flying, and how awareness constructs elements of knowledge for rules and exceptions. It is a work of theoretical value across disciplines and organizational settings and of practical importance as well. His lively narrative adds to translational research by translating knowledge or evidence into action in mission-critical systems.

Douglas V. Porpora
Professor of Sociology & Director
Communication, Culture and Media
Drexel University

Mission-critical systems are at the heart of NATO/military and any commercial aviation operations in which dedication to service, integrity and strength through diversity are the cornerstones. Utilizing this framework, Katerinakis explains the emergent knowledge elements in communication command with messages *"spoken-heard-understood-applied,"* from multiple stakeholders. Human factors, code-switching, organization, rule-following and exception-adjustment blend with the chain of interaction-awareness-knowledge. The interplay of theory and real-flight examples, with key interlocutors, creates a valuable narrative both for the expert reader and the lay-person interested in the insights of hospitals, nuclear plants, battlefields, safety and rescue systems, and crisis and disaster locations.

Ilias Panagopoulos
Command Fighter Pilot, Col (Ret)
Senior Trainer, Joint Aviation Authorities (JAA) Training Organisation
Safety Manager, NATO Airlift Management Programme

Foreword

I first had the pleasure of meeting Dr. Theodoros (Theo) Katerinakis virtually in January 2012 while serving on his dissertation committee at Drexel University through my duties as visiting research professor and aviation expert in residence. From my first look at his study proposal, it was clear he had a topic, and a way to present it, that could contribute not only to aviation safety but to the betterment of humankind. Each iteration of the manuscript reinforced my initial impression until the final draft in November 2014 prompted the following comments with my recommendation for acceptance of his dissertation:

> *Theo has offered the aviation community a wealth of knowledge and has added greatly to our understanding of silence, voice, and situation awareness in aviation safety. It would be wonderful if he would publish his findings as a "practitioner's paper" with the contributions summarized in the Conclusions section formatted for easy reading. In this way, he could "translate" this incredible work into information the operational community could put into practice. We owe Theo a world of thanks and respect for this work.*

Translational research as the art and science of bringing academic work and research study results to the operational world is my passion. I am thrilled that Theo continued on the journey to bring his work to you, the reader. There is irony in how communication is so important to sharing the information needed to build knowledge, though much information is never presented in a useful and usable format. Here we have an expert in communication theory who applies his communication knowledge to share his study results *in an understandable manner* for the benefit of all to use in their respective domains. Knowledge emerges through communication interaction with messages "*spoken-heard-understood-applied*," from multiple parties, while levels of awareness vary.

Communication is one of the most precious commodities for the creation and continuation of civilizations. Theo opens his book with the theory and scope of how communication is necessary for knowledge to form. Once the framework is built, he then masterfully moves to use aviation as an exemplar of the phenomena involved in the process of communication. Via a grounded theory approach, multiple techniques are used in the creation of *The Social Construction of Knowledge in Mission-Critical Environments: Lessons from the Flight Deck.*

Excellent literature reviews lead to summation statements that are supported by interviews, case studies, observations, protocol analysis, discourse analysis, and other investigative techniques. Each chapter brings to light a piece of the theory through enlightening examples.

One thread of interest to aviation practitioners is that all major human factors concepts taught to aviation personnel throughout the world (usually as a regulatory requirement) are given new meaning with the background discussions and applications to real-world situations. Although Chap. 2 focuses on human factors, the human factors theme is clearly evident on every page of the book. The conversational tone of the author's writing makes it easy to see how the academic construct impacts the performance of the actors in the aviation system.

Aviation is multinational, multicultural, and international in all aspects and is becoming more so day by day. This book has application for all actors in the worldwide system by investigating human behavior and human performance of the individuals and their intra- and inter-cockpit interactions. Whether it is pilot-to-pilot, pilot-to-cabin staff, pilot-to-air traffic controller, pilot-to-maintenance, manager-to-pilot, or any other communication combination, the process is consistent. There is a plethora of possible human and interaction errors, and this book teaches us how, why, and when communication plays a major part. The results suggest fruitful applications of theory to explore how knowledge is generated in highly structured, high-risk organizational environments, such as hospitals, nuclear plants, battlefields, safety and rescue systems, and crisis and disaster locations.

There are many "gems" in this book that are not found in other human factors or training manuals. For example, the discussion of the concept of "silence" and how it differs by culture, situation, stress, and training is something that we can all make use of inside and out of the aviation domain. The narrative unfolds knowledge advancement to expert status via the process of learning the "rule" and adjusting action to the "exception." Knowledgeable experts are distinct to know when and how rules are not enough to manage the situation. Then we get to a major topic of interest today – that of communicating with technology and automated systems. Human-automation teaming and the topic of trust – in humans or nonhumans – are woven skillfully throughout *The Social Construction of Knowledge in Mission-Critical Environments: Lessons from the Flight Deck*. This book's solid theoretical approach and excellent synthesis of the concepts supported by well-described examples make this an enjoyable educational journey.

Human Factors Scientist
US FAA Designated Pilot Examiner
Winter Haven, FL, USA
January 2017

Janeen Kochan

Janeen Adrion Kochan

Preface

This book was conceived as a research project to investigate the process of knowledge construction in critical environments, where humans interact and operate machines while taking complex decisions and actions with irreversible consequences. Aviation, ER medical care, and nuclear reactors operation are, among others, highly mediated critical environments. In aviation, human communication, safety, language security, pilot knowledge and expert performance, and a constructive approach to human error are knowledge factors that explain organizational behavior and management attributes.

The book connects communication with dynamic awareness (local, transitory, and global) and observes knowledge-required and knowledge-accumulated for successful operation. Knowing in an airplane flight follows the model "*spoken-heard-understood-applied*," as a communication session in the high complexity environment of the flight deck during the interaction of air traffic controllers and pilots. A flight according to flight operating manuals is the ideal situation; this book shifts hands-on knowledge for actual flight operations (the real flight situation) where the environment is more complex and the tasks for the crew (and controllers) have demands that compete for attention and interrupt one another in the course of time and with incoming signals. Both the crew members and the controllers need to monitor those tasks and pick up the thread to complete interrupted tasks and interactions.

The study conducted for this book identified topics of expertise, an instrument to elicit the knowledge and tools of experts, to represent their communicative behavior which conceptualizes their operation. Implementation was structured in the following stages: (i) realizing personal experiences in aviation communication from my logs, background, and visits; (ii) identifying knowledge topics that concern the aviation community and are important in accident investigation and the knowledge base of the industry, and start interviews; (iii) locating an applicable group of experts to forward data collection with in-depth interviews as discussions and few evaluators; (iv) developing an instrument with scenarios (refined from questions) to approach the experts in a more relevant manner; (v) articulating concepts, categories, and relations in a theoretical model for aviation safety; (vi) designing the methodological flow for the grounded theory (GT) approach to elicit and represent knowledge to evaluate the concepts using several iterations; and (vii) analyzing the collected data by selecting those that fit the theoretical model of aviation safety, by connecting the replies with knowledge construction using communication and expert performance theory. The first six stages are implemented in Chap. 4, with a complete, bottom-up, methodology for eliciting expertise, harvesting and representing knowledge. The second step has also used material in Chaps. 3 and 4 and the seventh step is covered on Chaps. 5 and 6. The two chapters cover knowledgeable sounds of silence, voice, and paralanguage with certain phraseology, using insights, testimonials, and commentation as narration-evaluation from the expert. Twenty-five aviation professionals (fighter and civil aviation pilots, flight engineers, accident investigators, and ICAO delegates) were located based on their expertise, safety culture background, and language sensitivity. They are not treated as subjects but as informants who offer their expertise. All responders are named from a major Greek airport name and narrate their insightful comments in a story-telling manner. Chapters 1 and 2 introduce communication concepts in aviation and connect human factors phenomena with knowledge in aviation.

Knowledge-observed, beyond applying the rule, is articulated when exception is identified: two landmark flights, unique in aviation history, and major scenarios and framing concepts for knowledge representation are explained in Chap. 3. Chapter 7 reviews several communication-oriented implications, points out policy issues, and discusses directions of further research.

Scenarios are described in 25 questions, sorted conceptually in 4 larger categories of interest, for in-depth interviews: communication phenomena (themes like silence and voice); the culture of professionalism (themes like relationship and familiarity); expertise, knowledge, and decision-making (themes like the two-challenge rule and unexpected events); and personal factors (themes like the perception on briefing, de-briefing, and oral information). The selection of guided scenarios for questioning was made so as not to satisfy preconceptions with predetermined replies but to make them relevant to what the participants experience in their daily routine and avoid "confirmation bias." The structure contains some control questions asking from the point of view of the "other side" of aviation professionals.

Knowledge construction negotiated in the book is based on communicative transactions and efficient representation techniques. Transactional communication determines the nature of the message as follows: (i) communication as an ongoing activity with its elements in constant change; (ii) a source is there only with a receiver and no message exists without a source (interdependency exists); (iii) communication is influenced by the individuals' interpretations, although they have a strong standardized foundation; and (iv) every participant is both a sender and a receiver. So, speaking, listening, confirming, and refusal to communicate is itself a mode of communication. There are at least four of those components that document the accomplished theoretical contributions of this book: (i) the deconstruction of silence phenomenon in multiple dimensions, as part of an interaction and accountability process; (ii) the synthetic proposition of a voice-categorical label which consolidates paralanguage and hesitation, nonverbal and verbal attributes, in a communication channel applicable in aviation; the complementary character (for the verbal or nonverbal message) of hesitations connects with different vocalic realizations – filling pauses, word lengthening, repetitions, and speech repairs; (iii) the incorporation of situation awareness, with its views of local, transitory, and global, as a component to transform communication to knowledge models, and (iv) the revisit of mother tongue as a non-conflictual but complementary communication tool which may facilitate linguistic security, as language carries knowledge.

The purpose of this book is to present the communication processes that construct pieces of knowledge and cultivate expertise in mission critical environments, where operators and interlocutors act according to what they communicate and communicate in order to act.

In the current "Innovation, Technology, and Knowledge Management" series, this book explores the impact of expertise at "micro" levels where teams and individuals apply their knowledge and affect the "meso" level (of the aviation industry), so that the "macro" level could adjust issues of safety culture and perception. Thus, aviation professionals and controllers will find in this book the expertise they like to communicate, students (in managements, human factors, psychology, human-machine interaction, aviation, etc.) will be able to comprehend knowledge data gathering and representation methodologies (mainly in information science and expertise inquiry), and the lay person will be able to evaluate the highly skilled knowledge workers in the aviation industry.

Aviate, navigate, communicate is the trifold of flight implementation in search of the flying instinct. The flow of knowledge in various awareness levels complements language instinct with the knowledge infrastructure of the reader.

Philadelphia, PA, USA Theodoros Katerinakis

Acknowledgments

I would like to acknowledge the support of my immediate family, the Springer series editor, a few close friends and research partners since my early college and later doctoral dissertation years for their patience, guidance and constructive approach during the process of completing the earlier research stages and the current project. There are several more specific contributions that I would like to point out from my scientific inquiries and considerable support.

Dr. Douglas Porpora, Professor of Sociology at Drexel, is acknowledged for the whole coordination, support, mediation and the open ear to discuss and refine all my ideas, as well as for prompting me to frame the broader project, and also for the numerous hours of scholarly (and not only) discussion which was quite helpful in keeping the focus for this research. My appreciation extends to his mentoring skills and his "broad-mindedness" to cope with diverse ideas, motivate and restrict when necessary, and being a real mentor for my academic personality in the USA.

Dr. Janeen Kochan is a Human Factors Scientist and Designated Pilot Examiner of "Aviation Research, Training, and Services, Inc." Her willingness to participate in this endeavor from the start of my research was really encouraging. Her systematic way of looking to my drafts and the comments on how to deal with pilots and controllers as "decision makers" and as problem-solvers and "managers" of unexpected events in a broad knowledge base, using concepts from grounded theory, were quite productive. Her remarks, as a content expert (both in the aviation field and the knowledge instruction field), offered a cleared orientation and several extensions of the current study. Furthermore, her availability to participate in this book with the foreword is a remarkable contribution.

Dr. Thomas Hewett, Professor Emeritus of Psychology and of Computer Science at Drexel, is acknowledged for his availability to read through my earlier texts, his steady pace to follow-up on the drafts as well as the accommodating approach during proposal defense. His directive to focus on knowledge acquisition from pilots, as "experts," was quite fruitful, as well as his suggestions regarding protocol analysis and some real-life cases. Also, his human-computer expertise was important for my approach to connect systems analysis with knowledge theory via human communication and operation automation.

Dr. Emmanuel Koku, Associate Professor of Communication at Drexel, is acknowledged for the time we have spent in discussions/classes of network theory and sociological issues of culture and power relations, as well as for his insights on research methods, power structures, and data collection. Dr. Koku was also supportive in one of my first relevant independent studies and other classes.

Dr. Alan Zemel, Associate Professor of Communication at State University of New York at Albany, is acknowledged for the infinite hours we spent in classes for conversation analysis, data collection, and suggestions on how transcripts are used. His detailed comments in the final draft of my research proposals were useful, mainly in the early phases of this study. Also, his feedback helped me investigating the key informants and appropriate methodology.

I would also like to thank Mr. Jack Medendorp, Executive Director of Human Research Protection of Drexel Institutional Review Board (IRB), for his guidance and cooperation and the opportunity to participate in the Collaborative Institutional Training Initiative (CITI). The training process was quite informative and constructive when I was able to complete it. This IRB level discussion provided useful insights for my preliminary research while interviewing my informants for the study at the Hellenic Civil Aviation Authority level.

Also, I would like to address special thanks to Professor Elias Carayannis for his recommendations and relevant material, his support during Springer evaluation process, as well as his systematic correspondence that kept me on top of the topic. His feedback on connecting communication discipline with knowledge theory was a catalytic direction to develop the current manuscript. I am quite grateful on his active engagement on this project.

I would also like to express my gratitude to the group of my informants/experts and the aviation community I have discussed my scenarios with, more specifically the members of Hellenic Civil Aviation Authority, the Hellenic Aviation Incidents Reporting Committee, the Hellenic Air Accident Investigation and Aviation Safety Board, the Special Committee for Air Traffic Management Terminology, the Control Tower shifts in "Eleftherios Venizelos" Airport of Athens, and the Area Control Center of Athens.

In the end, I am thankful to the community of pilots and operators of 115 Combat Wing (Chania, Crete, Greece) and US Naval Support Activity in Souda Bay (Crete, Greece) where I spent substantial time of my military service, as well as several pilots of the former Olympic Airways airliner. The actual names do not appear to preserve anonymity of my expert informants in the original research in HCAA prior to my Drexel University research. The blending of information theory and knowledge management has been cultivated through an open contact-line with major researchers in Athens University of Economics and Business, mainly the Department of Management Science and Technology where I was first introduced to systems analysis and information theory. Professor George Doukidis and long-time friend Dr. Katerina Pramatari were instrumental in maintaining my interest in business aspects of innovation and technology, knowledge acquisition, and expertise elicitation.

During the process, in my several trips back-and-forth between USA and Greece I received the support of my mother Vasiliki, as well as of my friends Sophia Georgiakaki in N. York, Dr. Andreas Karachristos and Dr. Laura Schleelein who hosted me during the crucial period of final text refinement in Philadelphia.

Lastly, but of crucial importance at a personal level, I am deeply thankful to my wife Maria and our son Antonios-Emmanuel who have supported me with the positive attitude during the writing endeavor. This book has a piece from each and both of them.

Contents

List of Abbreviations

115CW	115 Combat Wing, Chania, Crete, Greece
ACARS	Aircraft Communications Addressing and Reporting System
A-N-C	Aviate, Navigate, Communicate (A-N-C), is a phrase reminder of the pilot-in-command priorities during emergency situations and problem solving. Aviate – Maintain control of the aircraft, Navigate – Know where you are and where you intend to go, Communicate – Let someone know your plans and needs
ASRS	Aviation Safety Reporting System
ATC	Air Traffic Control, Air Traffic Controller
ATFM	Air Traffic Flow Management
ATM	Air Traffic Management
ATSB	Australian Transportation Security Board
AWACS	Airborne Warning and Control System (Aircrafts)
CDR	Cockpit Data Recorder
COC	Center of Operations/Communications
CRM	Cockpit Resource Management that evolved to Crew Resource Management
CW	Combat Wing
EAPA	Hellenic Aviation Incidents Reporting Committee (acronym in Greek semantics)
EASA	European Air Safety Agency
ELETO	Hellenic Society for Terminology (acronym in Greek semantics)
EUROCONTROL	European Organization for the Safety of Air Navigation
FBEA	French Civil Aviation Security Bureau (Bureau d'Enquêtes et d'Analyses pour la Sécurité de l'Aviation Civile)
FDR	Flight Voice Recorder
FIR	Flight Instruction Register (for air space control)
FOD	Flying Objects Debris
GA	Global (Situation) Awareness
GEETHA	Hellenic Joint Chiefs of Staff (acronym in Greek semantics)

GT	Grounded Theory
HAAIASB	Hellenic Air Accident Investigation and Aviation Safety Board
HAF	Hellenic Air Force
HCAA	Hellenic Civil Aviation Authority
HELLAS	GREECE (official name and known name of the country)
IATA	International Air Transport Association
ICAO	International Civil Aviation Organization
IFALPA	International Federation of Airline Pilots' Association
IFATCA	International Federation of Air Traffic Controllers' Associations
IFR	Instrument Flight Rules
LA	Local (Situation) Awareness
NASA	National Aeronautics and Space Administration
NATO	North Atlantic Treaty Organization
NTSB	(United States) National Transportation Security Board
OMEODEK	Special Committee for Air Traffic Management Terminology (of the Hellenic Civil Aviation Authority). An acronym in Greek semantics
RCAI	Romanian Civil Aviation Inspectorate
RoE	Rules of Engagement
SA	Situation Awareness (and Situational Awareness, in literature)
SATCOM	Satellite Communication (System)
SOP	Standard Operating Procedures
TA	Transitory (Situation) Awareness
TACAN	Tactical Air Navigation System
TRM	Team Resource Management
USN	United States Navy
VFR	Visual Flight Rules
VIP	Very Important Person

List of Figures

List of Tables

Part I
Background and Communication Phenomena in Aviation

Chapter 1
Introduction

1.1 Overview of the Current Work

In social sciences, social construction refers to the sociological approach of the nature of reality and has been associated with the postmodern era in qualitative research linked to the hyperbolic doubt posed by Bacon, the idea about how observations are an accurate reflection of the world that is being observed (Murphy et al. 1998). In constructivism, each individual mentally constructs the world of experience through cognitive processes, while social constructionism has a social rather than an individual focus (Young and Collin 2004). In this book, knowledge construction refers to how it emerges in a certain hierarchy from data and how it comes to have the significance for expert communities (of practice) and the broader society (Berger and Luckmann 1991). Such significance is realized in operating environments that support substantial infrastructure for society, as well as mission critical configurations which greatly affect the risk of human life.[1] The book focuses on transportation systems, and more specifically on flight realities, as a growing sector in world economy and society, a sector with remarkable safety record and prominent publicity in order to explain how knowledge is acquired and represented, as well as on the emergence of expert knowledge, developed by people devoting themselves to their subject (Berger and Luckmann 1991). In most cases, these experts claim novel status and

[1] In a formalized manner, the term "critical infrastructure" could be perceived as, namely, systems and assets, physical or virtual, so vital that the incapacity or destruction of such systems and assets would have a debilitating impact on security, national economic security, national public health or safety, or any combination of those matters (it is the meaning provided in Section 1016(e) of the USA Patriot Act of 2001 (42 USC 5195c(e)). Typically such type of infrastructure includes sectors like chemical and commercial facilities; communications; critical manufacturing and dams and emergency services; defense industrial base; energy and financial services; food and agriculture; government facilities; healthcare and public health; information technology; nuclear reactors, materials, and waste; transportation systems; and water and wastewater systems.

T. Katerinakis, *The Social Construction of Knowledge in Mission-Critical Environments*, Innovation, Technology, and Knowledge Management,
https://doi.org/10.1007/978-3-319-91014-7_1

ultimate jurisdiction over that type of knowledge, which this book is unpacking in a storytelling manner to unfold the layers of expert performance.

The social aspect accepts the view of society as existing both as objective and subjective reality. The social aspect accepts the view of society as existing both in objective and in subjective reality. Objective as realized through the interaction of people with the social world, so resulting to routinization and habitualization of their lifeworld, and consequently forming a general repository of knowledge. This is institutionalized at a societal level to the extent that future inquiries reflect to this type of knowledge as objective. Additionally, this objectivity is continuously reaffirmed in the individual's interaction with others and their corpora. Habitualization allows reproduction without much effort, while tacit dimensions of knowledge are case-based-sensitive and capture the exception. Taking the stance of the seminal Polanyi's work (1966) that people [or rather experts] know more than what they can tell, this book meets the work of Simon and Chase (1973) for perceiver-memorizer-problem solver and utilizes concepts for expertise and expert performer from the work of Ericsson (2006). In that sense it falls into the *Mode 3 Systems Approach for Knowledge Creation, Diffusion, and Use* in the conceptual profile of the current *Innovation, Technology, and Knowledge Management* series, described in Carayannis and Campbell (2009, 2010). Knowledge is negotiated in this book as acquisition, representation, and conversation with the expert doers and explores the micro and meso levels, in the aim of the series, for innovation to come.

The social construction of knowledge explained in this book takes the medium of language, as a key structural component, and builds on communication phenomena (like repetition, hear-and-read confirmation, and redundancy) of participant and co-participant actors. Language precedes and signifies concepts and provides a means of structuring the way the world is experienced. Language transmits thoughts and feelings and streamlines thought by constructing concepts. By advancing language, in the way experts use it, our problematic moves towards Pinker's idea of the human mind; the mind is a collection of instincts adapted to solve evolutionary significant problems (Pinker 1994). The expert decision-maker-operator in a critical environment of the flight is a perceiver of the constantly changing situation and the communication, in dynamic fraction of time with awareness. Conversation interactions of verbal and nonverbal communication are the most important carriers of maintaining, modifying, and reconstructing subjective reality in a shared objectified experience for all interlocutors-actors.

This book draws from the study of investigating how important human communication is in order to capture the knowledge level required to accomplish a safe flight or a fulfilled flight mission. Knowing comes from being aware in the form of understanding of what is happening and what could happen in the immediate and near future, as well as realizing the bigger picture of the flight conditions so as to accomplish a (safe) flight mission. The flight itself is a collaborative accomplishment in which participants operate with a certain knowledge background that is explicit and they construct different levels of knowledge in the social environment of shared meanings. They communicate in order to act, and they operate in a dynamic environment where *knowing means being aware and sharing awareness*

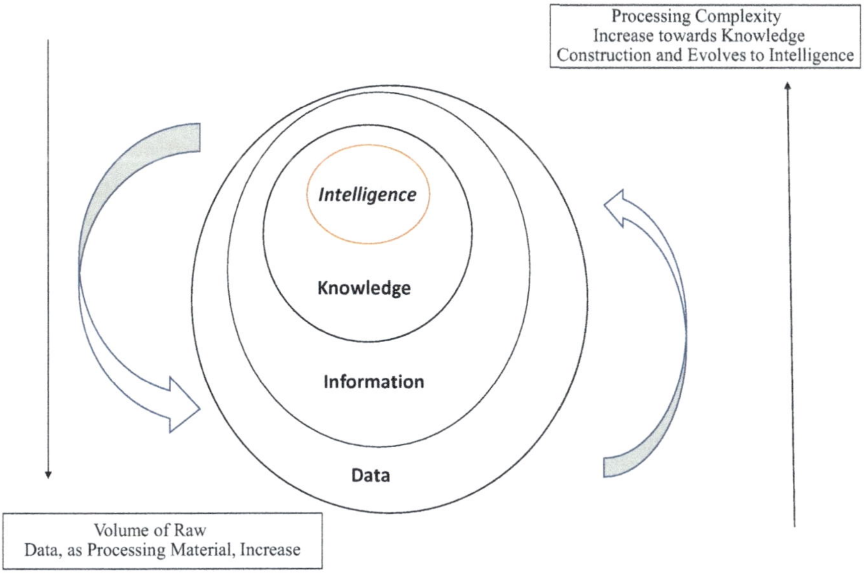

Fig. 1.1 The cycles of data transmission to encode-decode developing "knowledge (non-encoded)" to evolve for intelligence

with flight participants. That type of processing is represented in Fig. 1.1, in which input data are processed for information, and when combined together, knowledge is assembled, and its element is constructed for a result of intelligence.

The transmission rate and volume of data, as well as the operation complexity of the volatile aviation environment, are a depiction of the tight connection of communication with the type of knowledge constructed and applied. Moreover, knowing incorporates directing the flow of information for the current state of affairs from the expert performer (pilot or controller) to the mere operator (instruction executor), when the tacit knowledge allows the expert to identify exceptions, decide, operate, and communicate. In Davenport's[2] terms, participants in flight operation are considered "knowledge workers" in their own terms with the primary purpose on the creation, distribution, and application of knowledge.

[2] Tom Davenport's influential work on knowledge society, knowledge workers, and responses to automation is quite compatible with the knowledge-oriented operating flight deck, where human actors coordinate operations, experience their awareness, decide, act, and communicate in a highly mediated environment. One of his premises that "a lot of brain work including empathy – cannot be codified" connects the knowledge-working with the human actors. More on his work is available at http://www.tomdavenport.com/books/ and his special paper *Beyond Automation* at https://hbr.org/2015/06/beyond-automation. Davenport is the President's Distinguished Professor of Information Technology and Management at Babson College, the co-founder of the International Institute for Analytics, a Fellow of the MIT Initiative on the Digital Economy, and a Senior Advisor to Deloitte Analytics.

1.2 Personal Background Regarding This Study

I am not approaching aviation realities as a tabula rasa, as there are several preconceptions about the field of aviation, its position and perception in social life, the roles of individual actors in collective society, as well as the cultural aspects of the actors' behaviors. These preconceptions inform the content of several questions that were turned in scenarios for data collection (as shown in the Appendix A, B, and C). Certain conditions of human activity and experience imply some degree of predictability. In my case, my personal background in the Air Force, the bonding cultivated in the time of service in real crises, and the affiliation with several businesses active in military procurement due to my business involvement provide familiarity, access to people and sites, entrusted discussions, and valuation criteria.

Knowledge acquisition presented in this book uses a grounded theory approach to describe as well as explain, with the interpretive role of the researcher as facilitator from the inside[3] (Glaser and Strauss 1967). The description of my main experiences that follow is another indication of the density of time in the reality of life in Aviation and Air Force. Density is not only how we perceive a temporal process but how we identify the denseness of events in certain fractions of time. Although, I was on the ground service, the multiplicity of events, projects, missions, and service extends the perceptual time frame of the living experience.

1.2.1 Joining the Military After an Early Visit

My first experience with an Air Force base was on 1985 in 115 CW in Crete, Greece,[4] as a recipient of the high school prize in the regional mathematical contest. I have spent 3 days living in the guests' dorm and observed their typical routine activities, while I was able to spend 2 h in the flight simulator that was emulating

[3] In this study my approach is closer to the methodological school of Strauss in a hybrid manner with selective concepts from Glaser towards what works better for the data and the participants of the study. A grounded theory approach is not concerned solely to propose academic insights but also aims to show "why and how the theory can be used in practice."

[4] Souda Bay: Critical to NATO's Response. In 1990, the Hellenic Navy provided its naval base to the USA, creating NSA Souda Bay, which occupies about 110 acres and shares a home with the Hellenic Air Force's 115th Combat Wing and a Greek F-16 base. Hellenic Naval Bases are located on the north and south coasts of the Souda Bay harbor, and there is a civilian airport nearby in the city of Chania. NSA Souda Bay has supported US Air Force and Navy reconnaissance aircraft missions and other joint American and multinational operations. Currently, Souda Bay is the southernmost fighter base in NATO with one runway at the Hellenic Air Force's 115th Combat Wing, 31 Theater Air Base Vulnerability shelters and munitions storage. NSA Souda Bay is a designated Forward Operating Site, providing sustained support for deployed air and maritime operations to the USA, allies, and partners, as explained in Lexington's Institute report at http://www.lexingtoninstitute.org/wp-content/uploads/2016/03/Souda-Bay-NATOs-Military-Gem.pdf.

training flights. It was the first time I started keeping a journal, for nonroutine events, with an entry that "I should get back here when I join the military." Pages from that journal present the social and cultural environment which informs my participant observation for this study, from early on.

Coming from Crete, a Greek island with a pivotal geostrategic position, defense news and discussions are part of routine conversation in the public sphere. The long-standing tensions with Turkey rooted deeply in history, and the compulsory military services are significant parameters in the life of residents in the Greek island archipelago. Thus, most people in Greece feel that they have a right to address "their own military point of view," and the active service personnel (especially Special Forces and the Air Force) have developed a duty culture of service and defense motivation, which enhance professionalism. Unfolding the log of my journal, I could stop to the fact that the Air Force Academy is one of the most prestigious schools in the National University exam despite the news reports that pilots have an intense duty, high life risk, and isolation from social life while establishing camaraderie and brotherhood bonds between each other.

On 1995, I enlisted in the Air Force with a degree in information science and analysis; I was trained in HAF's General Staff as one of the five (only) information analysts of that year. After the first 3 months, I received advanced officer training in the Hellenic Pentagon being surrounded with "star-generals" insignia, not typical for airmen or other service officers who would meet and greet high-rank officers only in ceremonies. The overnight shifts I was assigned were either in the HAF's computer room (for the end-of-day information processing of HAF's General Staff till 4.30 a.m. with a Colonel of the Air Force) or in sentry points on top of the Pentagon wall with mixed patrols from the Joined Chiefs of Staff hierarchy. Both shifts were instrumental for my experience and familiarity with all aspects of military service. The intensity of those shifts led to a condition of exhaustion in the end of the service time, through which I was able to connect and spend time with the highly specialized medical personnel and officers of the 251 HAF's General Hospital. During my hospital treatment, I had numerous discussions with doctors who were evaluating physical and mental examinations for pilots and take several notes on perceptual and decision-making attributes.

After an evaluation of the first 2 intense months, I received the authorization of "Top Secret"; the upgrade came in the next 2 months to "Top Secret/Special Handling and NATO Confidential," the highest authorization for document handing and installation access in the military correspondence protocol. Thus, I was able to participate in security briefings, to monitor military cable signaling topics as new aircraft trial and delivery, military attachès transfers between embassies, and immediate delivery orders (meaning you receive the cable and leave your position to deliver hand to hand) from the Air Force Chief of Staff to the Commander of the Squadron to name a few of the topics. After a 6-month period and two consecutive evaluation exams, I was transferred back to 115 Combat Wing to become the associate in the Basic Operation Directorate of an Air Base with the top safety and targeting efficiency statistics in HAF.

Going back to the place I first visited on 1985, I found several civilian personnel members in supervisory positions (from subordinates at the time) and officers in charge that were my high school classmates at a rank of a flight lieutenant. Hence, I was able to participate in a demanding but accommodating atmosphere, like a member of an extended family. Soon, I started being involved with the operations of the Commanding Officer and the Deputy Commander both Colonels of HAF. My previous service record as well as my reporting capacity and diversity in service shifts (from flight object garbage collection to their children school bus escort, till overnight shifts in the Center of Operations and Communications) allowed me to cultivate a sense of trust, dependability, and mutuality, as well as to develop direct communication channels of immediacy with quite a few pilots, ground engineers, and commanding officers, with multiple overnight discussions about their readiness, the dangers they have experienced, the classmates they have lost in training, and their interpretation of the aviation instinct and ultimate goals.

1.2.2 A Crisis Experience and USN in Souda Bay

At the end of 1995 and the start of 1996, I was assigned to the Fighting Group in the COC of 115 CW during the Greek-Turkish crisis as a mission board presenter/notetaker. I have spent a 7-day week of consecutive 20-h shifts starting with national drill and continuing to the Imia crisis engagement team, with the 115CW Commander and 12 officers inside the operation center and all fighter pilots and engineers engaged in their hungers. I was responsible for the monitoring of the report back of the fighter jets, to stamp the cycle of "command-implement-report back," in a nonstop process where immediacy was more important than ranking. Tactical design deployments and weaponry configurations were allocated, in compliance with the "National Defense Plan" for retaliation/counterattack missions in the event of Turkish offensive activities in the Aegean Sea. The reference point for the territorial dispute was the Imia island complex called Limnia, close to island the island of Kalymnos (Katerinakis 2009).

The de-escalation of this crisis occurred by returning in the "status quo ante" (Ignatiou and Ellis 2009). Immediately after, I was transferred to the liaison position for HAF in the US Naval base in Crete. I was responsible for air traffic control clearances (for major freight aircrafts till fighter jets deployed for NATO or US missions), issuance of diplomatic identity cards and contractors' licenses, hunger security, and the fortnight reporting to the joined bilateral Foreign Affairs Committee. In that position, I had experienced numerous emergency landings, Airport Tower communication multiple times a day, cooperation with US Navy and US Air Force men and women during extended hours, security and patrol procedures, commander's briefings, coordination of logistics between 115 CW and Souda Bay, etc. I was able to get closer to the professional command-oriented culture in the US Armed Forces, the life balance of the personnel, and further cultivate my insights for air operations for NATO and US reconnaissance missions.

Furthermore, I have realized an important attribute in the value system of the US Navy/Air Force colleagues during our common shifts; their personality evaluation was based on a "job well done," a high-level of professionalism, and a reliable cooperation which cultivates dependability for trust. Although I was in an auditing position as a delegate of the HAF in USN activities, both their command and their personnel were quite expressive in their formal citation letters describing my duty, asking me to come in the base, as a civilian, and complete an operating manual for the liaison activities. These insights and the resultant familiarity or (sometimes) mutuality have informed several of my choices since then and shaped some of the questions in the scenarios, in a comparative mode between US and Greek military culture towards rules, safety, and efficiency.

1.3 A Flight: Knowledge and Control from A to B

A flight is accomplished when we can reach from point A to point B with safety and it operates in a communication continuum (air to ground and ground to air by means of communication devices, hardware indicators, and sensors). As you can see in Fig. 1.2, communication begins on the ground between cockpit and different types of controllers, depending on the altitude of the plane and the airport configuration.

In fact, communication starts (and maybe a crucial step) on the ground even before the planes start moving, even so with the preflight briefing, as well as with the visual inspection of the plane before getting in. And those conversations continue throughout with different control operators or teams. There are periods where silence is imposed and others when cockpit sterilization is required or communication is lost behind closed doors. Knowing how to and solving problems en route when needed are the major demands which become a knowledge accomplishment (consisting of the situation, intention, planning, historical log, conditions, and destination) when the flight arrives intact, with safety. Knowledge is identified in interpreting the application of the flight manuals and in reading the logs, as well as in problem-solving during the flight, in the relay of flight information via effective

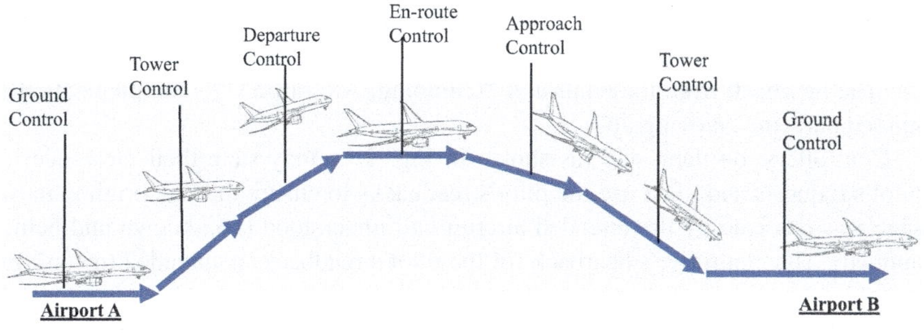

Fig. 1.2 Who has knowledge and control of the flight from "A" to "B"

processing, and with clear low-context communication of the realization of the flight. Another type of knowledge realization is what happens in interplane communication where interlocutors, sometimes, need to engage in nonverbal communication.

An airplane flight and its knowledge framework are represented, here, as an act of conversation that combines culture with communication of co-present, mediated, and cooperative actors. As a communication session, it is taking place in the highly mediated environment of the flight deck. The discursive space of a cockpit consists of actors that take roles and apply rules to complete tasks with their social actions. Conversation takes place in a time-critical situation, and all interlocutors must keep track of responding demands. They must understand the system in its entirety, position their role in it, and remain vigilant for consequentiality.

Talk and we will see silence too is inseparable from the task, as necessary to develop understanding of the flight situation. Language and action are juxtaposed. Talk and broader communication occurs not because people want just to interact but because they need to establish and sustain relationships for a cooperative environment, to get things done in their flight. Participants in flight situations are communicating (using verbal language but not only, as we explain in this study) not just to declare what is true or false but to "do things", correct errors, contribute to increase awareness for the other flight interlocutors, and report an action when done. Conversation utterances in flights are real-time speech acts. Interlocutors (crew members and air traffic controllers) act according to what they communicate and communicate in order to act. Even when interlocutors just report facts (the facts they observe in the situation), they meta-communicate the confirmation of what they have accomplished. When actions are attempted and not completed, communication is again the next step to continue.

1.3.1 Readback, Hearback, and Affordances

Pilots must respond to controller's instructions with a full readback of critical components of the instruction, in a cycle formulated in Fig. 1.3. For example, the message "Climb to 320, Aircraft 531" contains critical information that the readback "Roger, 320 for Aircraft 531" misses (the movement upward). With the missing part, the pilot does not understand whether or not the controller knows what is going on. The hearback from the controller "Climbing, Aircraft 531?" complements the missed part (the "moving up").

Controllers, on their side (as shown in Fig. 1.3), may state their clearance in pilot's requests and must use the pilot's readbacks to ensure that the original message was received by the intended aircraft and understood (i.e., known and being applied). The controller's hearback (of the pilot's readback) is intended to confirm that the pilot heard it right but also helps to validate that the controller send it right. This is the message-readback-hearback chain with the embedded redundancy which is typical during flight interactions. The language factor is crucial, as spontaneous

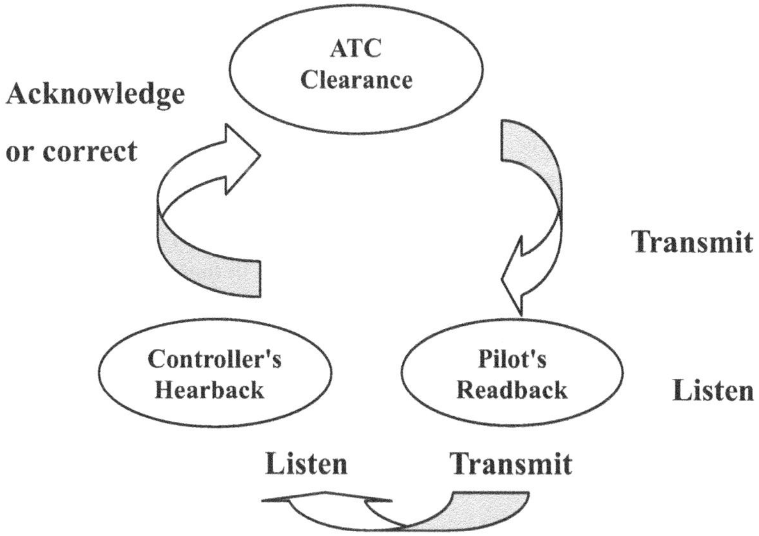

Fig. 1.3 Pilot-controller communication loop and model

interaction may deviate from the standard operating procedures (SOPs in flight manuals) in style or content (English or other, with code switching).

The communication model serves as the proper attainment within an ideal flight session from departure point to arrival point. Speakers and addressees work together to ensure that their utterances are mutually understood, as pilots are "many" to address "one" controller with accuracy and in a rapid pace. Controller-pilot communication follows a collaborative scheme in the form used for Fig. 1.3 (above). Aviation discourse is completely mediated, highly regulated, and intense to be mission critical, and primary actors (crew members and air traffic controllers –ATC) emerge from different organizational structures and cultures (Braithwaite 2001).

The discursive space of a cockpit includes actors, roles, rules, mediation equipment, and affordances as abilities to manage social interactions, emotional context, or social presence (Gibson 1979; Greeno 1994). The concept of affordances, as introduced by Gibson, would describe environmental attributes of the cockpit relative to human operators, who should possess the operating skills and the knowledge of decision-making. These are objects of the physical world in the cockpit, like information displays, control sticks, switches, buttons and paper materials of manuals, and the memory storage of flight recorders which enable or constrain behavior of their users. For the controllers, who are co-present in the cockpit, affordances describe their own operating environment in the ground control center with different information displays and handling and communication devices.

Gaver (1991) noted that affordances exist not just for individual action but also as possibilities offered by the physical environment for social interaction, including the affordance of functionality in the interaction between pilots and controllers. Greeno (1994) connected affordances with the abilities (as attributes) of the participants who contribute to the interaction. The cockpit and the ground control room relate the physical characteristics of information displays of pilots and controllers with the competence of both parties to communicate. In turn, inside the cockpit, affordances of functionality may determine the understanding of the interaction between pilot flying the airplane and copilot non-flying. The position of the side sticks, the mechanical connection of control wheels, the autopilot configuration, the colors of leds and switches, and the position of the crew rest area are all characteristics that connect perception with action inside the cockpit. The example of the 2009 crash of Air France AF 447 flight in the Atlantic Ocean is a typical occurrence of functionality-interaction-competence problematic. In that flight, the captain, the relief pilot flying (when captain was in rest), and the copilot could not afford to interact and act in their cockpit environment, when "all of a sudden everything started going wrong" as dictated in the accident investigation report (FBEA 2012). The notion of affordances is a significant part of knowledge factors collaborating to construct the necessary level of competence for a flight "mission accomplished."

1.3.2 Navigate (First), Aviate, Communicate

Following the simple phrase of "aviate, navigate, communicate,"[5] for a pilot in trouble, can mean the difference between life and death, as they are taught early on in their training. In terms of physical location of these communication sessions, "cockpit" or "control cabin" was used for the captain's station in boat transportation. In the late 1960s, when the Boeing 747 aircraft program was launched, the term "flight deck" was decided to include control cabin, extra crew seats, and a possible crew rest area with bunks. In this study, the term cockpit is used in reference of pilot-specific communication control; otherwise the two terms are used interchangeably. The marking of checklists for flight parameters, the collaboration of crew members inside the cockpit, and the report to air traffic controllers are greatly influenced by these functionalities. Furthermore, investigation or reenactment of incidents during accident investigations is based on such functionalities and the way they allow or constrain the action of crew members.

When flight crew members confront an emergency or abnormal situation in flight, they are expected to prioritize their immediate actions in the order of "aviate, navigate, and communicate" (A-N-C). "Aviate" means to ensure the safe flight path and condition of the aircraft (flying the aircraft but also the completing of checklists). "Navigate" means to find a safe flying position, decide to continue to, or divert from the originally intended destination or proceed to an emergency descent. The decision may be immediate but requires coordination via communication with air traffic control and other parties, except when carrying out critical activities that

[5]It is a very common phrase in flight schools and aviation safety instructions. Aviate, navigate, communicate (A-N-C) is a phrase reminder of the pilot-in-command priorities during emergency situations and problem-solving. Aviate – maintain control of the aircraft. Navigate – know where you are and where you intend to go. Communicate – let someone know your plans and needs.

will be reported afterwards (as it happened with the Falcon VIP flight in rapid descent). "Communicate" happens in the previous stages but may include a "standby" to save time or declare an emergency (when the pilots believe to be facing an emergency) as soon as possible (as noted in the example of Sect. 3.1.2 with the fatal delay) and cancel it later if normality is restored (as noted in Sect. 5.4.1 with controller's responses and expectations).

The cockpit (as the operating space to "navigate, aviate, communicate") in the flight deck is an automated cooperative environment and also a special case of a computer-supported collaborative environment (CSCE in Suchman et al. 1999; Suchman 1997, 2009). In its form of CSCE, interaction as attention management (cross-device interaction, cross-mode interaction), perception with comprehension, context management, and establishment and maintenance of relationships are instrumental. Human actors have to be capable of understanding the system as a whole and of positioning their role in it, identify with the group tasks and operation of automation parts, and remain conscious of *what differences their actions make*.

In this book, the exploration focuses on human communicative practices in various channels and tools employed from cockpit-to-cockpit, cockpit-to-cabin, cockpit-to-airport tower control room, and cockpit-to-command center (of operations in commercial flights or mission control in military flights). Communication interaction between human actors in those "spaces" establishes an integrative flow for physically present and co-present interlocutors. The modes, channels, practices, and concerns of their communication with the consequences to flight safety are in the core of this book, in order to highlight the catalytic role of communication in the knowledge constructed and used.

1.4 Communication Interaction and Outeraction

This communication framework moves forward to crisis interactions which are knowledge-demanding by nature, when flight situations contain nonroutine phases or are identified as flight in crises, as a whole. Therefore, we begin with two paradigmatic nonroutine flights that lay the foundation of several communication-oriented phenomena of interest aggregated under the labels of "silence" and "voice" (of Chaps. 5 and 6). The cockpit system has to be what engineers call a fault-tolerant system (in the event that a component fails, the faulty component is determined and taken off service, and a backup component or procedure can take its place immediately with no loss of service). This technical attribute impacts the way human actors operate in order to detect and correct errors and collaborate to ensure flight safety. But the only way to see that system functioning is in the signaling and ratification by speakers of work-arounds.

Literature on flights tends to focus only on conversation within the cockpit. I aim to extend this literature by incorporating, as well, the dialogue with the co-present interlocutors in the whole chain of interaction. The role of the air traffic controller as a substantial co-present actor is highlighted in this study. Furthermore, responders evaluate the instances when their interaction extends to "outeraction." Outeraction is a process of imminent conversation and flexible informal communication (Hwang et al. 2009; Nardi et al. 2000). In this view, outeraction is a catalyst for interaction, as social exchange is a preamble of the conversation for the main functions. Interaction, in this view, has a more specific meaning of information sharing and problem-solving. In the case of aviation, outeraction covers discourse that connects controllers- pilots with each other and manages their communication out of exchange information only. Controllers and pilots may converse with each other in social ways "outside" information exchange (using an alternative frequency, before or after the flight or even during the flight, in brief). Connecting in social ways when contextual features are negotiated through conversation facilitates information exchange. That kind of outeraction was also part of the introduction of my conversation with most of my aviation informants (as shown in the first stage conversations of Appendix C). Some informal conversation on mutual acquaintances in HAF and shared memories from mission design or air travel were useful to start the discussion and forward an interview to go deeper.

Psychologists and engineers studying flights have emphasized these issues of trust and safety, insofar, as they relate to technology and automation (Hopkin 1995; Muir 1987, 1994; Muir and Moray 1996; Muller 1996). In this book the literature is extended by researching interpersonal trust and what high-level knowledgeable practitioners call safety culture and how they interpret just culture and constructively understand human error and its consequences.

References

Berger, P., & Luckmann, T. (1991). *The social construction of reality*. London: Penguin Books.

Braithwaite, G. R. (2001). *Attitude or latitude? Australian aviation safety*. Aldershot: Ashgate Publishing.

Carayannis, E. G., & Campbell, D. F. J. (2009). "Mode 3" and "quadruple helix": Toward a 21st century fractal innovation ecosystem. *International Journal of Technology Management, 46*(3/4), 201–234.

Carayannis, E. G., & Campbell, D. F. J. (2010). Triple helix, quadruple helix and quintuple helix and how do knowledge, innovation and the environment relate to each other? A proposed framework for a trans-disciplinary analysis of sustainable development and social ecology. *International Journal of Social Ecology and Sustainable Development (IJSESD), 1*(1), 41–69.

Ericsson, K. A. (2006). The influence of experience and deliberate practice on the development of superior expert performance. In K. A. Ericsson, N. Charness, P. Feltovich, and R. R. Hoffman, R. R. (Eds.). *Cambridge handbook of expertise and expert performance* (pp. 685–706). Cambridge, UK: Cambridge University Press.

FBEA (Bureau d'Enquêtes et d'Analyses pour la sécurité de l'aviation civile, 2012). Final Report on the accident on 1st June 2009 to the Airbus A330-203 registered F-GZCP operated by Air France flight AF 447 Rio de Janeiro – Paris. BEA (French Civil Aviation Safety Investigation Authority). Paris, France.

Gaver, W. W. (1991). *Technology Affordances*, Paper in the 'CHI '91 Proceedings of the SIGCHI conference on Human factors in computing systems: Reaching through technology', ACM NY, USA.

Gibson, J. J. (1979). *The ecological approach to visual perception*. New York: Houghton Mifflin.
Glaser, B. G., & Strauss, A. (1967). *The discovery of grounded theory: Strategies for qualitative research*. New York: Aldine de Gruyter.
Greeno, J. G. (1994). Gibson's affordances. *Psychological Review, 101*(2), 336–342.
Hopkin, V. D. (1995). *Human factor in air traffic control*. London: Taylor & Francis.
Hwang, W. Y., Hsu, J. L., Tretiakov, A., Chou, H. W., & Lee, C. Y. (2009). Intra-action, interaction and Outeraction in blended learning environments. *Educational Technology & Society, 12*(2), 222–239.
Ignatiou, M., & Ellis, A. (2009). *Imia through the top secret US cables (ta Aporrita Tilegrafimata ton Amerinakon)*. Athens: Livanis Publication.
Katerinakis, T. (2009). Imia/Kardak crisis (1996): The night of grey compromises. In T. Katerinakis (Ed.), *MS portfolio in social networks, ethics, and international negotiations*. Unpublished Graduate Portfolio (pp. 35–65). Drexel iDEa E-repository at https://idea.library.drexel.edu/islandora/object/idea%3A3297. Accessed June 2014.
Muir, B. M. (1987). Trust between humans and machines; the design of decision aids. *International Journal of Man-Machine Studies, 27*, 527–539.
Muir, B. M. (1994). Trust in automation: Part I. Theoretical issues in the study of trust and human intervention in automated systems. *Ergonomics, 37*(11), 1905–1922.
Muir, B. M., & Moray, N. (1996). Trust in automation. Part II. Experimental studies of trust and human intervention in a process control simulation. *Ergonomics, 39*(3), 429–460.
Muller, G. (1996). Secure communication – Trust in technology or trust with technology? *Interdisciplinary Science Reviews, 21*, 336–347.
Murphy, E., Dingwall, R., & Greatbatch, P. P. (1998). Qualitative research methods in health technology assessment: A review of the literature. *Health Technology Assessment, 2*(16), iii–iix 1–274.
Nardi, B. A., Whittaker, S., & Bradner, E. (2000). Interaction and outeraction: Instant messaging in action. In *Proceedings of the 2000 ACM Conference on CSCW* (pp. 79–88).
Pinker, S. (1994). *The language instinct*. New York: Harper Perennial Modern Classics.
Polanyi, M. (1966/2009). *The tacit dimension*. Chicago: The University of Chicago Press.
Simon, H., & Chase, W. (1973). Skill in chess. *American Scientist, 61*(4), 394–403.
Suchman, L. (1997). Centers of coordination. A case and some themes. In L. Resnik, R. Saljo, C. Pontecorvo, & B. Burge (Eds.), *Discourse tools and reasoning. Essays on situated cognition*. Berlin: Springer Verlag.
Suchman, L. (2009). *Human–machine reconfigurations, plans and situated actions*. Cambridge: Cambridge University Press.
Suchman, L., Blomberg, J., Orr, J. E., & Trigg, R. (1999). Reconstructing Technology as Social Practice. *American Behavioral Scientist, 43*, 392–408.
Young, R., & Collin, A. (2004). Introduction: Constructivism and social constructionism in the career field. *Journal of Vocational Behaviour, 64*(3), 373–388.

Chapter 2
Communication and Human Factors Phenomena in Aviation Transmit Knowledge

2.1 Cockpit-Air Traffic Controller Communication

Silence and Voice Labels in Aviation Silence does not function in the same way in all cultures or subcultures. Silence can be attributed to low self-respect and self-consciousness in the professional environment. At other times there is an obligation of non-silence when reporting incidents or filling out an operational log. Humans are generally unwilling to speak up in organizational environments (Morrison and Milliken 2000, 2003; Milgram 1974). In flight communication team members in the airplane cockpit or cabin and sometimes the controllers function as a safety mechanism when they voice their opinion/observation in unusual or emergency situations. Besides voice, this book explores silence in personal, operational, institutional, and regulatory levels explaining what we know from monitoring silence and its impact on flight safety.

Silence is personal when a pilot is not talking or not reporting in flight logs as a choice, operational when it is needed for takeoff and landing, institutional from corporate culture and power structure, and regulatory when silence is imposed from regulations when emergency is happening. When silence is broken, a multitude of verbal and nonverbal signals are examined to explain how human communication works in flights where incidents occur, conditions of emergency are anticipated and then changed, or peak-point emergency escalates. Situation understanding as awareness, constraint time prioritization of speech and act, compulsory-embedded redundancy, and language security are concepts that emerge as contributions to communication theory and modeling, while flight safety is in top priority.

Cockpit-Air Traffic Controller Communication There are several critical phases in cockpit-air traffic controller communication cycle or continuum. Silence is part of the interaction, and voice should be used (till the computerized data link with text messages will be fully evaluated, accepted, and widespread). Controllers and pilots,

T. Katerinakis, *The Social Construction of Knowledge in Mission-Critical Environments*, Innovation, Technology, and Knowledge Management,
https://doi.org/10.1007/978-3-319-91014-7_2

using the vocal channel, assess the trustworthiness of each other (as speakers) and the information provided by the technology they use. By replacing voice with a (even high-quality) text message, the *manner, phrasing, and pacing of speech* is lost (Hopkin 1995).

Periods of silence are part of the process, but for how long, and what happens when silence should break with voice? Figure 2.1 (in two parts) breaks down a flight in stages, notes the switch between silence and voice in pilot-controller communication, and mentions the type of communication (part a). In part (b) the figure shows the percentage of flight accidents, in the flight stages of part (a) in order to connect communication with flight activities that have an impact on safety. Furthermore, part (b) includes the elapsed time (on average) for each stage of the flight as a percentage of the total flight time. Considering both parts of the figure in parallel, we can form a clearer picture about the type of communication, the relevant workload in the cockpit, and the time used for each flight stage. These aspects of a real flight situation provide insights about the stage when an accident/incident may occur. We will go through Fig. 2.1 with parts (a) and (b) in the following:

The intervals of part (a) show the type of messages; the stages of part (b) show the usual activity of the flight and the amount of time spent in that activity (per flight). For example, the cruising stage takes 60% of the time per flight; it is a period when 15.7% of the total number of accidents occurs, and all types of communication messages are exchanged. That is why the two parts are explained in parallel steps below. Silence and voice interchanges and variations determine *what we know* about the flight situation by affecting perception and expectation of the situation and by formulating awareness of the *here and now* in the flight.

In Fig. 2.1a a flight is accomplished when we can reach from a departure airport A to an arrival airport say B (in the direction of the dashed line on the bottom), from the left-hand side start to the right-hand side destination (in the bottom part of the diagram), with safety. The flight is divided to seven stages, three before and three after the main "axis" in the period of "all communication" in the center (it is a cruising phase of the flying aircraft, usually with the use of autopilot). The stages are represented with symmetry regarding what is communicated (from a regulatory point of view). The projected flight model is an average routine flight, approximating an ideal situation, as designed for flight instruction and training purposes in civil aviation. Those stages will be mentioned in the analysis of raw data, expert testimonies, and commentation of Chaps. 5 and 6.

In the start for part (a), doors close and cross-checked in the first stage on the left, in which only safety-related messages are communicated between pilot and controllers (*voice*). In part (b) it is the preflight taxing activity of ground movement for the aircraft. The second stage, with no contact messages (*silence*), has the ground acceleration for takeoff roll, the takeoff, and initial climbing till the landing gear is up. In part (b), it is the takeoff and initial climb stage of flying. In the third stage, with only safety-related messages to be communicated (restricted voice), the airplane is airborne and climbing till the seat belt sign is off (usually beyond the 20,000 feet, depending on the type of the airplane and the airliner). It is the climbing stage in

Voice- Silence Continuum and Stages of General Aviation Accidents Occurrences

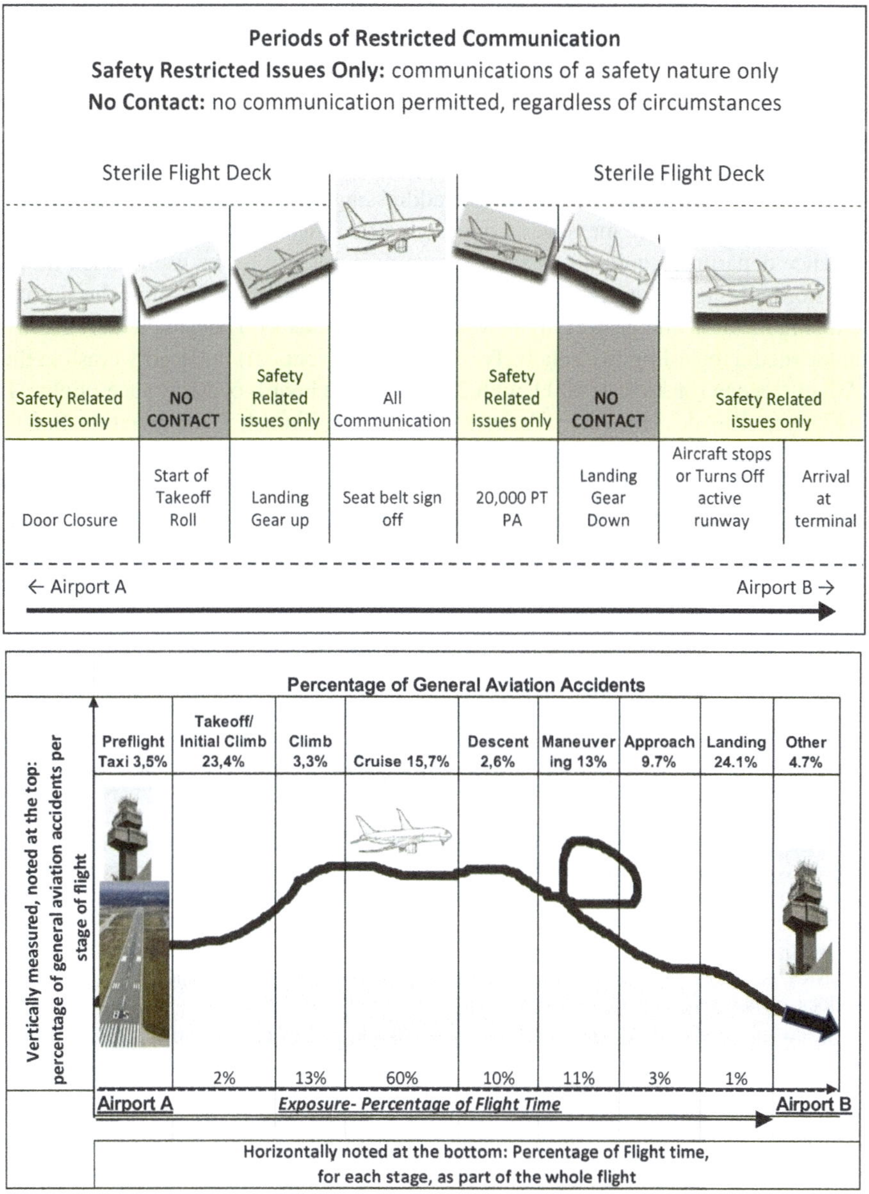

Fig. 2.1 (**a**) A flight as a chain of voice-silence periods (**b**) General aviation accidents per stage of flight

part (b). The fourth stage, open to all types of communication (*voice and silence*), is the stage of the "main" flight which corresponds to the cruising activity and usually covers 60% of the whole flight, in part (b). Although this stage is open to all types of communication and the use of autopilot mechanism, it is quite common in long flights over the oceans (and international airspace) to experience long periods with no vocal communication (similar to a communication vacuum) with the outside world (till the next ground-controlled airspace). Modern commercial aircrafts (after 1978) include an aircraft communication addressing and reporting system (ACARS[1]) to transmit and receive messages from ground facilities (airline operator and maintenance department, aircraft or system manufacturer, etc.). The use of such transmissions during the cruising stage (and beyond, in order to position the flight's cruising location and project flight systems "health checks") became widely known in the media regarding two relatively recent tragic events: (i) the deadly crash in the Atlantic ocean (on June 1, 2009, with 228 people on board) of Air France Flight AF 447 from Rio de Janeiro to Paris, for which the French Civil Aviation Safety Investigation Authority published a final report (FBEA 2012), and (ii) the disappearance of the Malaysia Airlines Flight MH370 from Kuala Lumpur to Beijing, on March 7, 2014, with 239 people on board, in the Indian Ocean (as a likely location) that is still under (unresolved) investigation.[2]

The fifth stage in Fig. 2.1a is the start of descent to the destination (part b) with only safety-related issues to be communicated (*restricted voice*). The flight is down around the 20,000 feet altitude, the seat belt sign is on, and there may be some maneuvering till the final approach to destination. Descent and maneuvering together could take 21% of the whole flight time (on average, in part b). The sixth stage is with no contact messages (*silence*), the landing gear is for landing. The aircraft lands, slows down, and turns off the active runway on the ground. It may

[1]ACARS can make use of the SATCOM, the radio system that uses satellites (for voice and data transmissions) to transmit its data to ground stations. The system rely on "pings" (like most utilities in Internet-connected machines), as simple probes used to check the reachability of SATCOM systems aboard the planes. The pings tell us that the plane has power and it is likely intact (because it has power). Eurocontrol's skybrary electronic repository provides a detailed description for the ACARS functions and messages at http://www.skybrary.aero/index.php/Aircraft_Communications,_Addressing_and_Reporting_System.

[2]The investigation for the disappeared Flight MH 370 is headed by the Australian Transport Safety Bureau (ATSB), at the request of the Malaysian Government. The government of Malaysia has officially declared the loss of Malaysia Airlines Flight 370 an accident, and all of its passengers and crew presumed dead, on January of 2015. Since January of 2017, ATSB has developed an operational reporting process to provide regular updates on the progress of the search effort for MH370 (in the special section at http://www.atsb.gov.au/mh370.aspx). At a meeting of Ministers from Malaysia, Australia, and the People's Republic of China held on July 2016, it was agreed that should the aircraft not be located in the current search area, and in the absence of credible new evidence leading to the identification of a specific location of the aircraft, the search would be suspended upon completion of the current search area. The statement said, "Despite every effort using the best science available, cutting edge technology, as well as modeling and advice from highly skilled professionals who are the best in their field, unfortunately, the search has not been able to locate the aircraft." So, search is now suspended but future hunt for missing plane is not ruled out.

take approach and landing time of part (b), with a total of 4% of the flight time. It is striking that this sixth stage accounts for 33.8% of the total number of accidents, in the top line of Fig. 2.1b. Lastly, in the seventh stage, the aircraft is taxing to arrive at the terminal, and only safety-related messages are communicated (*restricted voice*) in a ground movement that is not counted as flight time but still accounts for 4.7% of the general aviation accidents.

In Fig. 2.1a the first three and the last three stages correspond to the sterile cockpit implementation and compliance to general aviation regulations (and practices). The sterile cockpit/flight deck is the concept of restricting flight crew member activity to that "which is operationally essential during busy phases of flight" (Skybrary 2013). These phases are taxi out, take off, initial climb, intermediate and final approach, landing, and taxi in, as shown in Fig. 2.1a, b. Taxiing happens in two stages as the movement of an airplane on the surface of an airport, under its own power.

It is no coincidence that the no-contact periods (of silence) in part (a), when no communication is permitted regardless the circumstances, are the stages mentioned in part (b) that take only 6% of the flight time but account for the highest percentages of accidents (23.4% in takeoff, 33.8% in approach and landing), with the most workload for the crew. So, most of accidents occur when flights approach or depart from airports. In several studies (like NTSB 1994), it is indicated that maneuvering flight, approaches, takeoff/initial climb, and dealing with weather factors are flight activities with the likelihood of the most serious accidents.

When crew members are not focusing on their flight activities and engage in nonessential conversations or in activities totally unrelated to flying, critical information can be missed or misinterpreted (especially during the critical parts of the flight mentioned above). After reviewing several accident events, the US Federal Aviation Administration (FAA) established the explicit sterile cockpit rule in 1981, whereas in Europe the sterile cockpit concept is addressed by a European Commission Regulation in 2008.[3] Several of the aviation informants in the study for this book are addressing the sterile cockpit rule and practice mainly in Chap. 5.

2.2 Air Traffic Management

Considering air traffic controllers as another stakeholder, virtually co-present in the flight, the point of view of the controllers who participate in the study for this book approaches a flight as the object of air traffic management (ATM) as

[3] In the USA, FAA's Regulation #121.542 covers the topic and in Europe the European Commission Regulation EU-OPS 1.085 paragraph (f)(9)). The European regulation has a more mild phrasing compared to the explicit one in the FAA regulation: "The commander shall…not permit any crew member to perform any activity during take-off, initial climb, final approach and landing except those duties required for the safe operation of the aeroplane." More at http://www.skybrary.aero/index.php/Sterile_Flight_Deck.

implemented by Eurocontrol in Europe (Bonini 2005). ATM includes air traffic control (ATC) which deals with aircraft separation in the sky and the airport. Air traffic flow management (ATFM) is done before the flight with filing and analyzing a flight plan for each flight. In this process the expected capacity of the airspace is calculated, and the controllers dispatch turns with expected times for departures, for the safety of the flights and the economy of the airspace. When the capacity is reached, the aircraft must wait on the ground until it is safe to take off. Aeronautical Information Services (AIS) collects and distributes aeronautical information necessary to airspace users. These include information on safety, navigation, technical, administrative, or legal matters and their updates to synthesize the knowledge elements shared between pilots and controllers for each flight. In the study for this book, the work of the controllers is focused on ATC when they respond for their actions in real-flight scenarios.

Bonini's analogy for flight airways is the one of multistory motorways with two lanes, one lane from east to west and one lane for flights from west to east. The two lanes have a few different floors or levels. The civil controller needs to ensure that when the two lanes cross each other, aircrafts are safely separated, and, with an efficient flow. Regulation exists to avoid a conflict at crossing points of two or more aircrafts. The airspace of those lanes is divided in homocentric sectors around the airport with reference to their proximity and altitude. These sectors correspond to the stages of flight represented in Fig. 2.1. Also, the control exercised by the civil controllers follow the seven stages of the flight in Fig. 2.1a, b. From the ground at an airport upward, we have ground space (with ground control from gate to takeoff or from landing to gate), tower airspace (with tower control with initial climbing in departure or final descent in arrivals), terminal airspace (with departure control or arrival control, in a specified distance from airport A and airport B of Fig. 2.1), and en route airspace (with en route control, between departure terminal area and arrival terminal area). In several cases, oceanic control is included in the regulated spaces over the oceans. National airspaces in different countries have sectors, as volumes of airspace, which are crossed by many routes. The transition from one type of airspace to the other is a checkpoint in pilot-controller communication; e.g., in the route of the AF 447 flight before the deadly crash, the Dakar Tower (in Senegal) had no reply in its calls and notified Air France in Paris. Dakar Tower from Senegal was the expected checkpoint station call after Rio de Janeiro Tower in Brazil.

By controlling these airspaces and stages, the controller needs to detect conflict and then resolve conflict in airspaces by separating the flights in fractions of time. The controller is a decision-maker who must decide according to the demands of the operating environment, which may change after each decision action, as it happens to the pilots also. A flight is accomplished when we can reach from airport A to airport B with safety and operates in a communication continuum (air to ground and ground to air by means of communication devices, hardware indicators, and sensors). Communication begins on the ground between cockpit and different types of controllers, depending on the altitude of the plane and the airport configuration. In fact, communication starts (and maybe crucial) on the ground when the planes start moving. And those conversations continue throughout. There are periods where silence is imposed and others when cockpit sterilization is required or communication is lost behind closed (cockpit-locked) doors.

In the study conducted for this book, aviation informants share their stories which include narratives and experiences from flights in different airspaces and stages. For the fighter pilots, their flight is considered (here) as simplified to the same generic model, and their responses add a component of "airman" subculture with their mindset and camaraderie. That is why they are used in this study as expert performers (in the whole set of pilots), in comparison with the other pilots and actors, using Ericsson's terms (2006). Their controlled airspace includes several level of control (during airborne with the fighter formation chief who flies with them and the mission commanders (usually in two hierarchical levels) who coordinate what they do. The military controller is mainly dealing with the non-cruising stages of Fig. 1.3b. The sterile cockpit rule applies in the Air Force with takeoff and landing and *combat in* and *combat out* periods, during the cruising stage.

2.3 Human Factors Approach in Aviation

Human factors discipline considers communication and all the issues that influence human actors and technological systems performance, like job and roles, procedures and task design, team issues and groupware, as well as human machine interface. Human factors originates in psychology and focuses on areas like workplace safety, human error, product design, human capabilities, and human-computer interaction. It is often used in the USA synonymously with "ergonomics" that is the common term in Europe. Human factors is mostly cultivated in the US research tradition, whereas ergonomics is mostly concerned with design in the European tradition. Human factors, here, implicitly draws from systems theory.

Human Factors and Systems Human Factors and Ergonomics Society (HFES) suggests a broad definition that was initially adopted by the International Ergonomics Society (2000):

Human Factors or Ergonomics
Ergonomics (or human factors) is the scientific discipline concerned with the understanding of interactions among humans and other elements of a system, and the profession that applies theory, principles, data, and other methods to design in order to optimize human well-being and overall system performance.

Hence, human factors research incorporates actors but emphasizes on understanding of *how things get put together*, in a system of systems (Popper et al.

2004). So, human factors promotes a holistic approach, in which considerations of physical, cognitive, social, organizational, environmental, and other relevant factors are taken into account (Meister 1982, 2003). Human factors is involved with the creation of things and processes that effectively support people in their work and living activities (Lundin 1997). In the past two decades, there is a growing interest in the use of social science to examine ways that shape procedures and flow of processes, as well as tools and technologies used by people in real world and work settings (Heath and Luff 1991).

Going beyond physical human factors – dealing with human and biomechanical characteristics – communication interactions are more relevant with cognitive human factors, concerning with mental processes, such as perception, memory, and reasoning and, for the most part, with organizational human factors dealing with optimization of sociotechnical systems, in which organizational structures, policies, and processes are involved. Human factors could also inform synergetic activities in different fields of knowledge, such as telecommunications, informatics, electronics, and social science which lead to concepts like the Internet of things (Atzori et al. 2010).

Human actors are the protagonists in the organizational setting of human activity systems. We cannot solely address the reductionist approach of the parts of systems, but we must also achieve an integrated understanding of the operations of the systems themselves. The ultimate goal of human factors in aviation is to better match the system to the human and the human to the system (EATM 2007), although the human is part of the initial "system of systems." In this sense human factors is considered on the systems level (resources, design, and responsibility for system failure) and on the human performance level (awareness of the human actors, avoidance of overload for human actors, effect of stress and reliability, trustworthiness of the system). From an organizational perspective of human factors, dispersed work teams of liveware require social as well as technical support (Wellman et al. 1994; Wellman 1983, 1999; Garton et al. 1997).

Human factors work incorporates the holistic approach of examining the role of the human actor and human skills with technology, whereas systems analysis uses all systemic components that explain interactions, information generation/flow, and procedures. The cockpit system must be tolerant to function and accomplish the mission even when fault occurs. Most airline accidents are attributed to errors made by the flight crew (Loukopoulos et al. 2001, 2003, 2009). Uncovering the causes of such an error is one of the greatest challenges in aviation operation and accident investigation, as many factors intermingle. That is why a flight mission should be represented via its communication occurrences, as interactions that materialize the intangible session of a flight. This communication point of view enriches the typical causality that searches for a "human error," with informative interactions and oral documentation.

Communication has long been suggested as a critical issue in all aspects of human interaction. Most testimonies of ASRS and National Transportation Security Board database illustrate how critical communication is in aviation and aviation safety, from the cockpit-controller interface to coordination in the cockpit to cockpit-cabin interaction to the management of safety and creation of a safety culture (Krivonos 2007). Communication has a multifaceted nature with a variety of settings and situations. Kanki and Palmer (1993) provide a useful structure on how communication functions in aviation and aviation safety. Specifically, communication affects crew performance by providing information, establishing interpersonal relationships and predictable behavior patterns, maintaining attention to task and monitoring, and providing a management tool for crew resource management (CRM) (Helmreich et al. 1999; Helmreich 1997; Helmreich and Foushee 1993). Hence, in this book, human factors converges with a systems approach with emphasis in coordination and collaboration. Apart from that, analysis is focused on informational conversation acts performed by interlocutors who participate in the system operation.

Aviation Discourse Communication flow operates the system even in cases of increased sophistication. Engineers and pilots as two distinct communities in NASA's Langley sites had a troubling communication to assess whether design changes helped when articulating the problem of aircrafts, as closed mechanical systems or as open human activity systems (Vincenti 1993). Communication in human factors research is a projection of all essential considerations in making technological systems safe, effective, easy, and enjoyable to use and aviation systems to fly and land on destination under certain conditions. Identification of cognitive, task, and organizational factors that make human actors vulnerable to error can be accomplished in their conversational interactions. Symptoms of miscommunication which include violation of procedures and instructions as well as communication-specific problems and phraseology defects are disclosed using suggestions from conversation analysis approaches.

Human factors analysis is focusing more on conditions and evaluations, whereas in cockpit operation environment, the issue seems to be more on how pilot, crew, and ATC expertise is to be exercised and (thus) implemented in a dynamic decision-making process. Pilots and ATC need to decide, re-decide, mutually confirm, and continue in their turn-taking, and only their conversation is showing *how they do that*. Thus, communication features like corrections of previous messages, evidence of misunderstanding, hesitation, hedges, emotional words, acknowledgments, affirmations, status reports, reports of reports, reports of action, and reports of status are needed to investigate the interactional organization of cockpit activities (Krifka et al. 2003).

Cockpit operation through conversation is understood as a process concerned with the reconstruction of segments of information and not with psychological interpretation (only). The epistemology of systems thinking applied to human activity with emergence, hierarchy, communication, and control paves the way to analyze aviation discourse to answer the question of *how interlocutors understand each other, ensure that they do so, and trust each other*. This type of understanding and trust-building is synonymous to operational knowledge.

2.4 Language-Based Communication Errors in Aviation

In aviation, communication occurs in an environment which challenges effectiveness and where failures may have severe consequences. What people claim to know is realized in their communication – for although English as a lingua franca was designated for international aviation use in 1951 (Rubenbauer 2009), substantial issues of culture (in terms of habits and safety), professional behavior, operational training, and compliance in standardization still exist (Seidlhofer 2005).

As Cushing (1994) has analyzed, contextual constraints and language are combined in communication problematic inside the flight deck and when pilots and controllers interact. Jones (2003), in his review, writes about "symptoms of miscommunication"; those include procedure or instruction violation (like clearance avoidance) and communication-specific problems (like arbitrary interpretation, dialects, and inappropriate phrasing). Also, Jones speaks about "phraseology defects," like multiplicity of meaning, multiplicity of synonyms, and implausible words, with a contextual meaning that may be different from the colloquial meaning.

Cockpit conversation generates meaning through detailed order of spoken exchange in which rules, machinery, and structures manifest themselves in specifiable forms. Cockpit conversation is human communicative behavior with elements like those identified by Lorenz (1973):

Aviation Speech Elements Interaction
Spoken not yet heard; heard not yet understood;
understood not yet agreed;
agreed not yet applied;
and applied not yet always applied

Incident and accident data from flight conversation concern units of speech sent but not yet received on the other end and utterances received but not yet perceived from both ends and when perceived may not yet read back, as mutual confirmation or dependability on both parts develops. Even when instructions are read back, there is a step forward needed to manage and coordinate acceptance in language for international multilingual flights.

Moreover, the margin of error is large, and considerable amount of redundancy is designed into the system (Leveson et al. 2002). An example of redundancy in the ATC system is the pilot's reading-back to the controller instructions just given (by the controller) to confirm that the instruction has been "heard and understood by the pilot in the flight concerned." The units and elements of information provided by a pilot is not always a sole hearback communicated by the controller. Beyond that, the information elements communicated by the pilot is compared with the actual aircraft's trajectory on the controller's radar screen, to ascertain that the instruction has been followed.

Conversation order is occasioned and meaningful for those who produce the respective utterances. Speakers and addressees need to be aware of each other, where they are located (demographic understanding), what the others know (knowledge understanding), and what they are able to do (capability understanding), in order to complete a communication transaction. Preliminary interviews for this study revealed the transactional mode in conversation in which interlocutors in the cockpit (corporeal present or co-present) establish a relationship of trusting partners when their conversation escalates towards crisis warning phrases (Watzlawick and Beavin 1967; Schegloff 1982). It is a case of people-to-automation trust and people-to-people trust (Lee and See 2004), in which partnerships may project overreliance on automation (Sparaco 1995), misuse or disuse of automation (Panasuraman and Riley 1997), or failure of crew intervention (NTSB 1997) in which knowledge is being constructed and disseminated on the go. Trusting the others and the available technology, at the appropriate level, is regarded as an effective strategy to manage workload.

The language factor is crucial, as spontaneous interaction may deviate from the standard operating procedures (SOPs in flight manuals) and rules of engagement (RoE, mainly in military aviation) in style or content (English or other, with code switching) of its tacit knowledge. Analysis of genre of interviews will reveal aspects of how cockpit conversation with ATC involved is a staged, goal-oriented, and purposeful activity (Martin 1984).

In this context, there are several communicative problems that arise. One set of problems involves issues of reference, repetition, ambiguity, sequence breaking, and the like (Cushing 1988, 1995, 1997). A second set of problems concerns the implementation of SOPs (and RoE in military aviation). Although there are highly detailed SOPs, what counts as following them in a crisis situation is not always clear, and sometimes it happens not to be followed at all. These problems need to be negotiated in time-critical situations, as long as trust in automation and among interlocutors is established.

In this book we examine how this is done; how, for example, pilots and air traffic controllers talk, interact, and outeract with orderliness; how they engage in error monitoring and correction for conversation repair; and how they negotiate whose interpretation of an SOP or implementation on RoE will prevail. Also, in the process, it will be examined how a trusting environment is fostered and how inner and outer identities and culture are sustained in crisis conversation.

Research topics are investigated around the following overarching inquiry: How is it possible for interlocutory actors to understand each other, ensure that they do, and establish trust to rely to each other and their equipment? Figure 2.2 describes the effect of various communication errors in aviation safety, due to the incomplete knowledge coming from erroneous awareness. The effect to aviation safety comes through the label of a "non-desirable situation," less severe like the "incident" or crucial like "an accident." In any case it is a troubling effect, as represented by the "airplane in a knot."

In message exchange, ATC and pilots speak to each other by addressing the "receiver first" unlike ordinary conversation between people where they address "the sender first," both with a call sign (like "JFK Tower" or "US 531"). Pilots and controllers exchange strings of information and imperative sentences and are trained

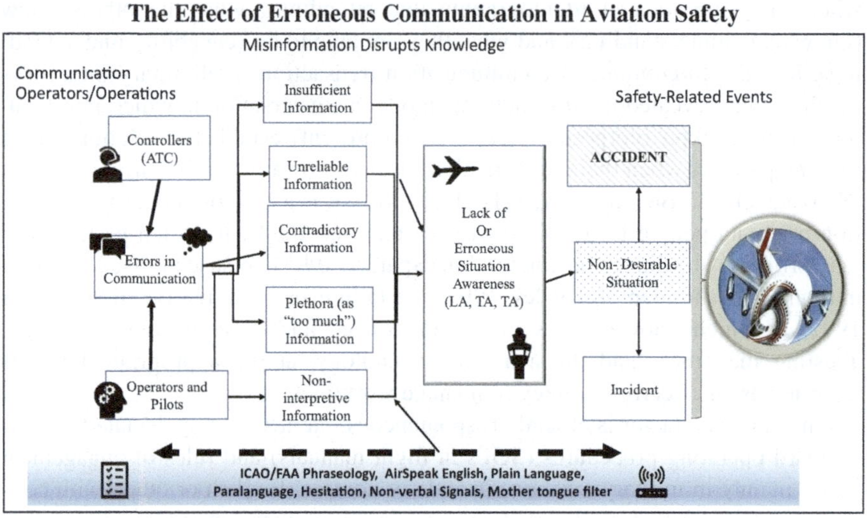

Fig. 2.2 Communication errors and aviation safety events

to remember and implement those strings and then report back. A string is typically a unit of information containing a parameter and a value (for the parameter). The sentence *turn left to three five zero* has the instruction (turn left) and a given value (three five zero). That is a single unit of information.

Howard (2008) extends the single unit to a complex unit of information: "five three oscar alpha, turn left heading one three zero, go down and maintain ten thousand, contact arrival one thirty-nine point two." This message contains four units of information: (i) call sign of the hearer "53 OA," (ii) turn left with a numerical value (130), (iii) go down and numerical value (10,000) to maintain, and (iv) contact another controller (arrival) with the numerical value of the frequency (139.2). This definition of units of information shows that not only the number of words and the time are used to verbalize but also the informational content of each element that may accumulate workload (Barshi 1998).

Many inquiries will arise as such data are examined regarding aviation safety and how the message is propagated in order to be received; e.g., what is the appropriate number of units of information per turn (in the exchange between controller and crew)? Is it recommended to use unitary information or be economical on the frequency and contain multiple units per interaction? Is there a need of follow-up, especially here when a readback is quite complex and the resultant hearback maybe further complex? Is it safer to have one unit per exchange, similarly to one word-one meaning instruction, or it is not realistic? The quantity and quality of information mentioned in Fig. 2.2 is directly relevant with those exchanges of units of information. Safety relates with the designed redundancy through overlapping expertise of staff, mutual checking (between crew members inside the cockpit, between controllers and pilots, and between controllers working together), and technological support to capture information from different sources and locate erroneous messages or actions.

Figure 2.2 starts on the left with the top and bottom box of communication actors. The major participants in aviation communication are the controllers and the pilots (with the airline operators who also participate and interfere in the discussion, usually from another frequency). In these inputs (from operators), we could include messages rooted by mechanical devices (like the autopilot or information displays) in the cockpit or the control room, but the role of automation in human communication is in the side of this study. Their communication may generate errors in the way both parties interpret the message to extract information.

The erroneous content of these messages is divided to five categories (in the boxes in the middle). First is insufficient information when, for example, a prompted action or a call sign or an instruction or a direction number is missing. Second is unreliable information when, for example, hesitation mutes a given instruction or when the prompt has uncertainty modifiers. Third is contradictory information, for example, a message or a reply or a hearback in contrast with the weather, altitude, direction, or a previous transmitted message. The fourth box has a plethora of information that is rather "too much," meaning inappropriate for the situation and the timing transmitted. Relevant examples would be, when several instructions are prompted in sequence instead of one-by-one with proper timing or when the sterile cockpit rule is broken by a controller or crew member. The fifth box contains non-interpretive information, like a simple "OK" that cannot connect with a call, an incomplete hearback, and the use of nonstandard words or words from a non-shared mother tongue.

2.5 Situation Awareness and Communicative Implications

All these symptoms of erroneous communication led the arrow to the box of erroneous situation awareness, the knowledge initiator in this case. Thus, errors are coming in and an erroneous result is generated. Situation(al) awareness (SA) emerges and is analyzed as a key concept of this book. In terms of the "Erroneous SA" (as perception) is box of Fig. 2.2, it refers to the pilot's (and the controller's in several cases) mental model for the evolving tasks to execute for the situation. SA includes the acquisition representation, interpretation, and utilization of information to stay in control. It is the comprehension of the operational environment and other situated factors affecting current and near future goals. SA enables or restricts, if erroneous, the rapid decision-making, effective action, and the communication of those. The control of complex, dynamic systems and high-risk situations like those happening in flights goes side by side with the level of awareness. The erroneous SA that may start from the pilot (and transition to the controller) means that the pilot does not know "what is going on" and (therefore) "cannot figure what to do." The lack of SA, due to erroneous inputs, is a primary factor in accidents attributed to human error. Erroneous SA leads to the generic non-desirable situation with two types that vary in severity. Incidents are usually violations of instructions or legal parameters, like bypassing a clearance, violating aircraft separations, not filing a flight plan, etc. Incidents might still affect operational safety in aviation, as lessons used to avoid accidents. Unlike the perception of accidents in aviation where the situation is more important than the blame, discourse about accidents, as a category of events, has several conceptual parameters to consider. Suchman (1961) has tightened the criteria of labeling an "incident" (quite severe or even fatal) as an "accident" with several factors: expectedness (unanticipated or not), avoidability (preventable or not), intentionality (intended as outcome), degree of warning (no warning leads to accident occurrence), the duration of occurrence (quick phenomena are easily labeled as such), the degree of negligence (if negligence is associated, it is less likely to be an accident), and the degree of misjudgment (more misjudgment is less likely to call it an accident). In any case, the accident is mostly labeled because of its consequences in human lives or material destructions (Vigilant and Williamson 2003). It is considered comforting, a consolation to survivors and protection of the victims to speak about an accident. In Chap. 3, we will use the discourse about specific accidents in the specific context of aviation and human error patterns and resolution. In Chap. 4, the knowledge elements discovered are represented with the use of a grounded theory approach.

The last part noted in the Fig. 2.2 is the foundation box in the bottom which contains the floating communication tools in human aviation communication (with the dashed arrow in both sides of the diagram). Phraseology used is standard from International Civil Aviation Organization (ICAO) or US Federal Aviation Administration (FAA). Despite the desired level of standardization, discrepancies in the use of typical terms in English phraseology still exist between ICAO and FAA, as exemplified in Appendix D. Another aspect of language is the mix of plain English language with technical jargons and colloquialisms and the nonverbal phenomena of paralanguage and hesitation. Nonverbal signals are used in problematic situation to complement the verbal language. Lastly, the mother tongue cultural component affects all information exchanges, especially when details are needed or an emergency is happening.

2.6 Implementing the Knowledge Study with Aviation Professionals

The study conducted for this book had to identify topics of expertise, an instrument to elicit the knowledge and tools of experts, to represent their communicative behavior which conceptualizes their operation. Implementation was structured in the following stages: (i) realizing personal experiences in aviation communication from my logs, background, and visits; (ii) identifying topics that concern the aviation community and are important in accident investigation and the knowledge base of the industry and start interviews; (iii) locating an applicable group of experts to forward data collection with in-depth interviews as discussions and few evaluators; (iv) developing an instrument with scenarios (refined from questions) to approach the experts in a more relevant manner; (v) articulating concepts, categories, and relations in a theoretical model for aviation safety; (vi) designing the methodological flow for the grounded theory (GT) approach to elicit and represent knowledge to evaluate the concepts using several iterations; and (vii) analyzing the collected data by selecting those that fit the theoretical model of aviation safety, by connecting the replies with important topics in communication theory. The first six stages are implemented in Chap. 4. The second step has also used material in Chaps. 3 and 4, and the seventh step is covered on Chaps. 5 and 6.

With a series of visits, I was able to locate aviation safety phenomena of organizational and institutional interests. I wanted to evaluate several experiences realized in the Air Force and actively observe in aviation installations, control and command shifts, and briefing-debriefing activities. A process of constant comparison starts with the first level of reading and identifying phenomena of interest, and then a decision on the theoretical sampling is taken. Twenty-five aviation professionals (fighter and civil aviation pilots, flight engineers, accident investigators, and ICAO delegates) were located based on their expertise, safety culture, and language sensitivity. They are not treated as subjects but as informants who offer their expertise. More details about them are documented in Appendix E. All responders are named with the convention *aviation informant professional number 1 to 25*, a list of which abbreviated with their alias name from a major airport in Greece[4] is shown in Fig. 2.3.

Scenarios are described in 25 questions, sorted conceptually in 4 larger categories of interest, for in-depth interviews: communication phenomena (themes like silence); the culture of professionalism (themes like relationship and familiarity); expertise, knowledge, and decision-making (themes like the two-challenge rule); and personal factors (themes like the perception on briefing-debriefing). The selection of guided scenarios for questioning was made so as not to satisfy preconceptions with predetermined replies but to make them relevant to what the participants experience in their daily routine. It is a step necessary to avoid "confirmation bias" (Lewicka 1998). Thus, scenarios are grounded on anonymous testimonies of aviation professionals, in national reporting systems (usually in a confessing style) or in briefing (rather debriefing) remarks of empirical situations as aviation professionals experienced them. The flow of questions which elaborate the scenarios is sequential, but responders are aware of the routing of the questions that explain the expected point of view of the answers. The structure contains some control questions asking from the point of view of the "other side" of aviation professionals. The full list and wording of the scenarios is included in Appendix A.

The replies and commentary in the scenarios were sorted to match the theoretical model of aviation safety developed on Sects. 4.3.1 and 4.3.3 (Fig. 4.2). Comparisons among different responses were made to relate replies with categories of interest, during an iterative process. Further analytical sorting was accomplished through

[4] Greece has a remarkable density of airports in its sovereign territory: 46 airports in an area size of 131,957 km^2 in a coastline of 15,000 Km, with a very strong aviation tradition and operation. Also, Hellenic Air Force, although founded in a defense doctrine and strategy, is a protagonist in most NATO and international aviation drills and evaluations. The history of Greek Aviation starts at 1911.

1.	Eleftherios Venizelos -AIP1, Fighter Pilot
2.	Ioannis Daskalogiannis- AIP2, Fighter Pilot
3.	Thessalonikis Makedonias AIP3, Fighter Pilot
4.	Nikos Kazantzakis - AIP4, Fighter Pilot
5.	Ioannis Kapodistrias - AIP5, Fighter Pilot
6.	Ippokratos Koos - AIP6, Helicopter pilot
7.	Democritos Alexandroupolis - AIP7, Fighter Pilot
8.	Osysseas Elytis - AIP8, Civil Aviation pilot
9.	Ifaistos Limnios - AIP9, Civil Aviation pilot
10.	Diagoras Rodios - AIP10, Civil Aviation Pilot
11.	Dionysios Solomos -AIP11, ATC
12.	Aristarchos Samios - AIP12, ICAO delegate
13.	Megas Alexandros -AIP13, Accident Investigator
14.	Ikaria Ikaros -AIP14, ATC
15.	Oia Santorinis -AI1P5, ATC
16.	Vasileios Pyrros -AIP16, ATC
17.	Omiros Chios -AIP17, ATC
18.	Filippos Kozanis -AIP18, ATC
19.	Aristotelis Kastorias -AIP19, ATC
20.	Alexandros Onassis -AIP20, Flight Engineer
21.	Dimitrios Vikelas -AIP21, ATC
22.	Vassilis Konstantakopoulos -AIP22, ATC
23.	Neos Aghialos -AIP23, Fighter Pilot
24.	Palios Ellinikos -AIP24, Civil Aviation pilot
25.	Stan Dardman -AIP25, Civil Aviation Pilot

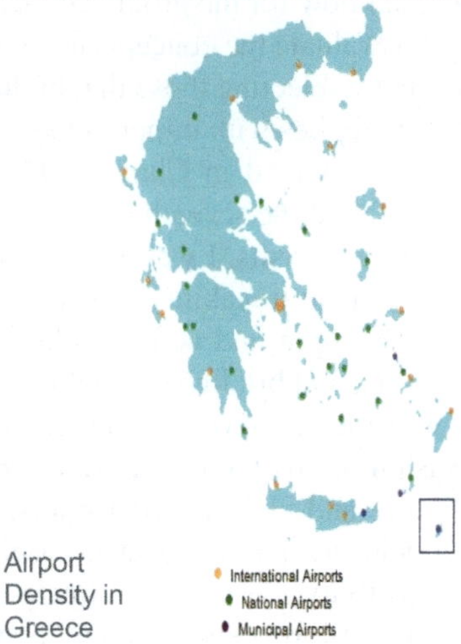

Fig. 2.3 Naming convention table of informants in Greek Airport mapping

coding the textual information with labels of action and interactions. In this sense, Chaps. 5 and 6 connect the instances recorded with the communication concepts of theoretical significance and compatibility with the core concept of aviation safety.

Contributions to Communicative Knowledge Components Little is known about the human communication use of silence and voice in flights, especially in unusual or emergency situations. In addition to voice, this study explores the role of silence (personal, operational, institutional, and regulatory) and its impact on effective flight communication. In aviation communication interaction is purposeful, since pilots and controllers develop consciousness of *where one is and where is the other and in what status* only when they exchange messages. The channel of voice between pilot and controller may contain periods of operating in silence, but voice should restart in order to have a meaningful exchange of information (with no uncertainty) between their physically distant spaces. The empirical data from the aviation informants used in this study are expected to accommodate a whole range of instances of voice and silence. The channel of voice has a scope covering verbal signals with certain phraseology till truncated messages of hesitation, interrupted messages, and dialogic marking of checklists.

In communication terms airplane flights are abstract representations of the basic model of communication with flight operating manuals dictating all activities; but this rather happens in routine cases as ideal flight situations. The routine cases are very important as they frame a sense of "flight normality" and assign attributes and expectations to the human participants involved. Schemes of conversation from actual flights and incidents will be used to examine how communication is produced and how each recipient acts on that. Such conversations in textual forms are common in archives of Aviation Accident Investigation Authorities (like ATSB, USNTSB, EASA, FBEA, and HAAISB; also see Walters and Sumwalt 2000; MacPherson 1998, Murphy 1980). Different scholars had highlighted diverse aspects in such texts like linguistic factors (Cookson 2009; Nevile and Rendle-Short 2007), problematic discourse (Howard 2008), message misunderstandings (Barshi 1997), and human error and expertise (Dismukes et al. 2007). The current study aims to show the convergence of that conversation in communication terms.

Lastly, from a methodological point of view, the incorporation of a grounded theory (GT) approach in this study resonates with the people who experience the phenomena under study and have a professional interest on the study (pilots, air traffic controllers, and accident investigators (Brown et al. 2002; Strauss and Corbin 1990)). The GT approach is used to elicit and represent the knowledge elements identified in the complexity and variability of human action in the mission critical environment of a flight where communication and action goes hand-by-hand. This type of communication inquiry in aviation connects "what is really going" with data from the field to ratify the belief that humans take an active role in responding to problematic situations and are sensitive to the evolving events in microscales of time and show the relevance of theory grounded in data. The observed interrelationships in aviation among conditions (structure), action (process), and consequences are quite compatible with the communication continuum necessary for the flight operation. This book is a realization that meaning is defined and redefined through interaction and the human participants act, based on meaning and the semantics of their knowledge.

References

Atzori, L., Iera, A., & Morabito, G. (2010). The internet of things: A survey. *Computer Networks, 54*(15), 2787–2805.

Barshi, I. (1997). *The effects of linguistic properties and message length on misunderstandings in aviation communication*. Unpublished Ph.D. dissertation in Linguistics, University of Colorado, Boulder.

Barshi, I. (1998). *The effects of mental representation on performance in a navigation task*. Ph.D. dissertation in Cognitive Psychology, University of Colorado, Boulder.

Bonini, D. (2005). *A Model of Trust in the work of Air Traffic Controller*. Unpublished Doctoral Dissertation, EUROCONTROL Experimental Centre (EEC), EU, Brussels.

Brown, S. C., Stevens, R. A., Troiano, P. F., & Schneider, M. K. (2002). Exploring complex phenomena: Grounded theory in student affairs research. *Journal of College Student Development, 43*, 173–183.

Cookson, S. (2009). Zagreb and Tenerife: Airline accidents involving linguistic factors. *Australian Review of Applied Linguistics, 32*(3), 22.1–22.14.

Cushing, S. (1988). Language and communication-related problems of aviation safety. Washington, DC: U.S. Department of Education. (ERIC Document Reproduction Service No. ED 296595, FL 017 504.).

Cushing, S. (1994). *Fatal words: communication clashes and aircraft crashes*. Chicago: The University of Chicago Press.

Cushing, S. (1995). Pilot-air traffic control communications: It's not (only) what you say, it's how you say it. *Flight Safety Digest, 14*(7), 1–10.

Cushing, S. (1997). Language differences in aviation communication: Problems and solutions. In P. Quigley & P. McElwain (Eds.), *Proceedings of the Aviation Communication: A Multi-Cultural Forum*.

Dismukes, R. K., Berman, B. A., & Loukopoulos, L. D. (2007). *The limits of expertise: Rethinking pilot error and the causes of airline accidents*. London: Ashgate Publishing.

EATM (European Air Traffic Management). (2007). The human factors case: Guidance for human factors integration. Eurocontrol Paper Online. Available http://www.eurocontrol.int/humanfactors/public/standard_page/HF_Case.html. Accessed 3 Nov 2011.

Ericsson, K. A. (2006). The influence of experience and deliberate practice on the development of superior expert performance. In K. A. Ericsson, N. Charness, P. Feltovich, R. R. Hoffman, & R. R. (Eds.), *Cambridge handbook of expertise and expert performance* (pp. 685–706). Cambridge: Cambridge University Press.

FBEA (Bureau d'Enquêtes et d'Analyses pour la sécurité de l'aviation civile 2012). (2012). Final Report on the accident on 1st June 2009 to the Airbus A330-203 registered F-GZCP operated by Air France flight AF 447 Rio de Janeiro - Paris. BEA (French Civil Aviation Safety Investigation Authority). Paris, France.

Garton, L., Haythornthwaite, C., & Wellman, B. (1997, June). Studying online social networks. *Journal of Computer-Mediated Communication, 3*(1). http://jcmc.indiana.edu/vol3/issue1/garton.html [Reprinted in S. Jones (Ed.) (1999). Doing Internet Research (pp. 75–105). Thousand Oaks, CA: Sage.].

Heath, C., & Paul, L. (1991). *Work, interaction and technology: Empirical studies of social ergonomics* (Technical Report EPC-1991-108). Taylor & Francis Group Ltd. Reprinted with kind permission, Proceedings of the 11th International Conference on Ergonomics, July 1991, Paris, France, Taylor & Francis Group Ltd., London.

Helmreich, R. L. (1997). Managing human error in aviation. *Scientific American, 276*, 62–67.

Helmreich, R. L., & Foushee, H. C. (1993). Why crew resource management? Empirical and theoretical bases of human factors training in aviation. In E. Wiener, B. Kanki, & R. Helmreich (Eds.), *Cockpit Resource Management* (pp. 3–45). San Diego: Academic Press.

Helmreich, R. L., Merritt, A. C., & Wilhelm, J. A. (1999). The evolution of crew resource management training in commercial aviation. *International Journal of Aviation Psychology, 9*(1), 19–32.

Hopkin, V. D. (1995). *Human factor in air traffic control*. London: Taylor & Francis.

Howard, J. W., III. (2008). 'Tower, am I cleared to land?': Problematic communication in aviation discourse. *Human Communication Research, 34*(3), 370–391.

International Ergonomics Society. (2000). What is ergonomics the definitions 'The discipline of ergnomics'. Online HTTP: Available at http://www.iea.cc/01_what/What%20is%20 Ergonomics.html. Accessed 6 Oct 2011.

Jones, R. K. (2003). Miscommunication between pilots and air traffic control, 2003. *Language Problems and Language Planning, 27*(3), 233–248.

Kanki, B. G., & Palmer, M. T. (1993). Communication and crew, resource management. In E. Wiener, B. Kanki, & R. Helmreich (Eds.), *Cockpit resource management* (pp. 99–136). San Diego: Academic Press.

Krifka, M., Martens, S., & Schwarz, F. (2003). Group interaction in the cockpit: Some linguistic factors. In R. Dietrich (Ed.), *Communication in high risk environments* (pp. 75–101). Hamburg: Helmut Buske Verlag.

Krivonos, P. D. (2007). Communication in aviation safety: Lessons learned and lessons required, regional seminar of the Australia and New Zealand Societies of Air Safety Investigators 9–10 June 2007.

Lee, J. D. & See, K. A. (2004, Spring). Trust in Automation: Designing for appropriate reliance. *Human Factors, 46*(1), 50–80.

Leveson, N., de Villepin, M., Srinivasan, J., Daouk Neogi, N., Bachelder, E., Bellingham, J., Pilon, N., & Flynn, G. (2002). A safety and human-centered approach to developing new air traffic management tools. In *Proceedings of the 4th USA/Europe Air Traffic Management Seminar*, 3–7th December 2001, Santa Fe, USA.

Lewicka, M. (1998). Confirmation Bias: Cognitive error or adaptive strategy of action control? In M. Kofta, G. Weary, & G. Sedek (Eds.), *Personal control in action: Cognitive and motivational mechanisms* (pp. 233–255). New York: Springer.

Lorenz, K. (1973). 'Nobel lecture video: Analogy as a source of knowledge'. Nobelprize.org. December 12, 1973. Online. Available : http://www.nobelprize.org/mediaplayer/index.php?id=1582&view=1 (Accessed November 3, 2011).

Loukopoulos, L. D., Dismukes, R. K., & Barshi, I. (2001). Cockpit interruptions and distractions: A line observation study. In R. Jensen (Ed.), *Proceedings of the 11th International Symposium on Aviation Psychology*. Columbus: Ohio State University Press.

Loukopoulos, L. D., Dismukes, R. K., & Barshi, I. (2003). Concurrent task demands in the cockpit: Challenges and vulnerabilities in routine flight operations. In *Proceedings of the 12th International Symposium on Aviation Psychology* (pp. 737–742). Dayton: The Wright State University.

Loukopoulos, L. D., Dismukes, R. K., & Barshi, I. (2009). *The multitasking myth: Handling complexity in real-world operations*. Farnham: Ashgate.

Lundin, R. (1997). Human factors and interactive communication technologies. *Canadian Journal of Educational Communication, 26*(2), 61–74.

MacPherson, M. (1998). *The black box: All new cockpit voice recorder accounts of in-flight accidents*. New York: William Morrow & Company.

Martin, J. R. (1984). Language Register and Genre. In F. Christie (Ed.), *Language studies: Children's writing: Reader* (pp. 21–29). Geelong: Vic Deakin University press Reprinted with revisions in A. Burns and C. Coffin (eds), (2001). Analyzing english in a global context. London: Routledge, pp. 149–66.

Meister, D. (1982, June). The role of human factors in system development. *Applied Ergonomics, 13*(2), 119–124.

Meister, D. (2003). *Conceptual foundations of human factors measurement*. Abingdon: Taylor & Francis.

Milgram, S. (1974). *Obedience to authority*. New York: Harper & Row.

Morrison, E. W., & Milliken, F. (2000). Organizational silence: A barrier to change and development in a pluralistic world. *The Academy of Management Review, 25*(4), 706–725.

Morrison, E. W., & Milliken, F. (2003). Speaking up, remaining silent: The dynamics of voice and silence in organizations. *Journal of Management Studies, 40*(6), 1353–1358.

Murphy, M. (1980). Review of aircraft accidents. In G. E. Cooper, M. D. White, & J. K. Lauber (Eds.), *Resource Management on the Flight deck: Proceedings of a NASA/Industry Workshop* (NASA CP-2120). Moffett Field: NASA-Ames Research Centre.

National Transportation Safety Board (NTSB). (1994). *Safety study. A review of flight crew-involved major accidents of U.S. air carriers, 1878 through 1990*. Washington, DC: Department of Commerce, National Technical Information Service.

Nevile, M., & Rendle-Short, J. (2007). Language as action. *Australian Review of Applied Linguistics* (special thematic issue Language as Action: Australian Studies in Conversation Analysis, edited by Rendle-Short, Johanna; Nevile, Maurice), *30*(3), 30.1–30.

NTSB (National Transportation Safety Board). (1997). *Marine accident report – Grounding of the Panamanian passenger ship Royal Majesty on Rose and Crown Shoal near Nantucket, Massachusetts, June 10, 1995 (NTSB/MAR97/01)*. Washington, DC: NTSB.

Parasuraman, R., & Riley, V. (1997). Humans and automation: Use, misuse, disuse, abuse. *Human Factors, 39*(2), 230–253.

Popper, S., Bankes, S., Callaway, R., & De Laurentis, D. (2004, July 21–22). *System-of-Systems Symposium: Report on a Summer Conversation*. Arlington: Potomac Institute for Policy Studies.

Rubenbauer, F. (2009). *Linguistics and flight safety: Aspects of oral English communication in aviation*. Aachen, Germany: Shaker.

Schegloff, E. A. (1982). Discourse as an interactional achievement: Some uses of 'uh huh' and other things that come between sentences. In D. Tannen (Ed.), *Analyzing discourse: Text and talk* (pp. 71–93). Washington, DC: Georgetown University Press.

Seidlhofer, B. (2005). English as a lingua franca. *ELT Journal, 59*(4), 339–341.

Skybrary. (2013). *Sterile Flight Deck (HF- AO)*. Eurocontrol Electronic Repository/Human Factors/Flight Safety. Online. Available HTTP http://www.skybrary.aero/index.php/Sterile_Cockpit, Stored June 17, 2013. Accessed 11 June 2014.

Sparaco, P. (1995). Airbus seeks to keep pilot, new technology in harmony. *Aviation Week and Space Technology, 142*, 62–63.

Strauss, A. L., & Corbin, J. (1990). *Basics of qualitative research. Grounded theory procedures and techniques*. Newbury Park: Sage.

Suchman, E. (1961, August). A conceptual analysis of the accident phenomenon. *Social Problems, 8*, 241–253.

Vigilant, L., & Williamson, J. (2003). To die, by mistake: Accidental deaths. In *Handbook of death & dying* (pp. 211–223). Thousand Oaks: SAGE Publications.

Vincenti, W. G. (1993). *What engineers know and how they know it: Analytical studies from aeronautical history*. Baltimore: Johns Hopkins University Press.

Walters, J., & Sumwalt, R. (2000). *Aircraft accident analysis: Final reports*. New York: McGraw-Hill Professional.

Watzlawick, P., & Beavin, J. (1967). Some formal aspects of communication. *American Behavioral Scientist, 10*, 4–8.

Wellman, B. (1983). Network analysis: Some basic principles. *Sociological Theory, 1*, 155–200.

Wellman, B. (Ed.). (1999). *Networks in the Global Village: Life in contemporary communities networks in the Global Village*. Boulder: Westview Press.

Wellman B. S., Salaff, J., Dimitrova, D., & Garton, L. (1994). *The virtual reality of virtual organizations*. Presented at American Sociological Association, Los Angeles.

Part II
Flights, Scenarios, and Knowledge Representation

Chapter 3
The Falcon, the Helios, Two Scenarios, and Framework

3.1 Two Landmark Flights

Aviation is clearly the safest form of transportation. That safety comes, in part, from the particular attention paid for each and every flight, any complain, report, and incident and the volume of knowledge accumulated from all stakeholders in the aviation community, including passengers. Typical mottos in the industry convey messages like *we have people, who check the people, who check the people, who check your aircraft* to emphasize accountability and safety based on the accumulation of knowledge layers from people, manuals, and information processing from observation, operation, and function. Along these lines, nonroutine cases that go beyond the routine flights are spotted for "more lessons learned." There is a consistent and continuous feedback mechanism maintained in the industry: Each flight has a flight log to report, and the report is considered as "raw data" for the manufacturer. Thus, attention to detail goes hand in hand with attention to exceptional situations.

The two flights explained in this chapter were selected as landmark in two ways: (a) In the Falcon VIP flight of 1999, the Deputy Foreign Affairs Minister of Greece with his son and other passengers were deadly injured while the aircraft was shaking as an "afraid horse," but the pilot flying managed to land/save the aircraft (and himself and those who were buckled) that had done things no other plane had survived in aviation history, and (b) the Helios 522 "ghost plane" on 2005 is a case whose flight path and condition are followed from the outside and not from the inside (by the intercepting fighter jet pilots, with very limited voice recording data), as everyone (except one person) in the plane was unconscious. Both flights were selected to highlight the process of social construction of knowledge as negotiated by the immediate actors/participants, as well as by the accident investigators/committees who were involved to resolve the questions asked from the industry and the community at large. Both flights possess unique attributes in

T. Katerinakis, *The Social Construction of Knowledge in Mission-Critical Environments*, Innovation, Technology, and Knowledge Management,
https://doi.org/10.1007/978-3-319-91014-7_3

the history of accident investigation worldwide and qualify for what is described as "out of the rule," falling in the category of exceptions and contributing to specific lessons learned for aviation community at large.

The accident investigation of the Helios flight concluded with a reenactment of the actual flight, as if the bottom-up grounded theory approach was part of the actual investigation (with backtracking of the events). The Helios 522 accident investigation from the Hellenic Air Accident Investigation and Aviation Safety Board (HAAIASB) validated its findings with an additional reenactment of the accident flight designed "by the textbook" (HAAIASB 2006) to apply a hindsight mechanism. An identical Boeing 737 was reserved using the same flight parameters and the same schedule and route from Larnaca (Cyprus), with the hypoxia conditions and pressurization switch to the "manual" position, the interception with F-16 fighter planes, etc. It took off carrying only the HAAIASB committee members, the district attorney, and paramedical personnel. The new flight was conducted in the actual (not simulated) environment and confirmed the findings (Tsolakis 2013).

In both cases, after the events a series of recommendations were implemented in several agents and tools of the sector: the operators, the manufacturers, the flight and maintenance manuals, the training topics, and setting of controllers (at a national level for Greece, international level from ICAO and FAA).

3.1.1 A Major Accident in the Saved Falcon 900B SX-ECH VIP Flight

The so-called afraid horse VIP flight,[1] with a Falcon 900B SX-ECH, which reshaped the concept of silence in terms of accountability in the eyes and ears of the aviation community, is the flight of the VIP aircraft of the Greek State on September 14, 1999. The pilot of that flight, namely, *Ifaistos Limnios*, has witnessed a major incident of rapid descent without controls, and he was able to hold the shaking aircraft and to prevent it from crashing. The story is narrated retrospectively after spending days in face-to-face conversation combined with multiple secondary sources, as well as a reality check from the accident investigator.

The airplane had a flight log with a repeated unresolved entry regarding the maximum speed and the autopilot use, as well as a repeated warning signal in the cockpit that needed further investigation although not crucial for flight navigation. Ifaistos Limnios did not accept institutional inertia of not speaking up when a repeated remark is entered in the flight log. He kept entering the comment as soon as he was entering the aircraft. The aircraft cabin defect log is usually occupied by the cabin crew members, and the aircraft technical log is occupied by the flight deck members. The two logs combined communicate flight realities to the next shifts, whereas the two flight recorders – for voice and indications – are the real-time

[1] This Falcon flight was labeled as "shaking" by the media or as "afraid horse" or "coffee shaker" in the minds of the survivals because of the feeling they got in the rapid descend (with ups and downs) and the effort to stabilize altitude with pilot's manual virtuosic efforts.

archives of the flight. Also, he did not want a copilot to remain silent (in a VIP flight) and always chooses to fly together with the specific fellow pilot who understood the speaking up mindset from their briefing. He always wanted to be in the aircraft at least 2 h before flight to perform his preflight ritual: Walk around the aircraft for a visual check. Ifaistos Limnios was only imposing the sterile cockpit rule in his take-off and landing, a practice he was keeping religiously, regardless the VIP passenger and the status of the cabin.

This flight created a completely new framework in discussing silence, as a parameter of employment relations and personal accountability. The flight is remembered by HCAA personnel as a road map in corporate governance in the aviation industry and repositioned aviation safety as a lasting responsibility before and after the flight. It was a VIP flight in the sense of an official state flight with a governmental diplomatic delegation.

Ifaistos Limnios was an experienced and accomplished pilot who spent a career in the state-owned airliner Olympic Airways. Working as a jet pilot was a dream that came true for Ifaistos Limnios, as he was growing up when Olympic Airways was privatized by the late iconic tycoon Aristotle Onassis and became an international player in the industry during the 1970s. He was selected to lead the Prime Minister's crew, although without political affiliation, due to his decisive cruising style and his open character in crew teamwork (as he noted in the discussion).

What Happened in the Falcon Flight The flight took off in a bright clear day of the Greek fall. The Aviation Safety Network[2] summarizes the notable part of the flight when approaching to Romania in the extract below:

> The Falcon reached a cruising altitude of FL400 until 47 minutes from take-off, when a normal descent to FL150 was initiated, with the autopilot engaged in vertical speed mode. Approaching FL150, the co-pilot had a request for a further descent. Just before FL150 the ATC re-cleared the flight to continue descent to FL50. One second later the autopilot disengaged and thereafter the aircraft was manually flown by the captain. Between FL150 and FL140, for approximately 24 seconds, the aircraft experienced 10 oscillations in rapid descent which exceeded the limit of the load factor and all controls in the cockpit did not respond. At the request of the flight crew, radar vectoring was provided by the ATC, and a visual approach and landing was performed on runway 08R. It appeared that the cabin interior had been completely destroyed, resulting in fatal injuries to seven passengers.

The passage epitomizes the Falcon 900B SX-ECH[3] flight which started as an ideal flight situation with cruising autopilot, in a bright shining day, although the

[2]The Aviation Safety Network is a private, independent initiative founded in 1996. Online since January 1996, the Aviation Safety Network covers accidents and safety issues with regard to airliners, military transport planes, and corporate jets. The ASN Safety Database contains detailed descriptions of over 10,700 incidents, hijackings and accidents (as described at http://aviation-safety.net/about/).

[3]It is a French manufactured plane by Dassault Aviation, the company founded by the pioneer industrialist Marcel Dassault immediately after WWII. It holds a wide market share for executive business jet or jet-taxi for VIP passengers (or those who want to project the idea of a very important person). Several models are now in the market and are considered to be in the elite of private jets, as explained at http://www.dassaultfalcon.com/en/Pages/Home.aspx

aircraft had a history of malfunctions[4] that were not addressed promptly by the manufacturer,[5] as Ifaistos Limnios points out. The whole process is a continuum and requires appropriate input from the pilot-in-command to communicate it to the airliner-operator which has to communicate the maintenance difficulties to the original manufacturer (like Dassault in this case). The footnoted safety enhancement SE170 accents that communication flow and exchange from these parties are not efficient, as Ifaistos Limnios realized during his service. He started the chain of events and the sequence of reporting multiple times, but the expected actions were not taken.

Back to the flight from Athens to Bucharest, Romania, the aircraft experienced ten perturbations in the axis of the wings, for approximately 24 s, which exceeded the limit maneuvering load factor. The descent was in a high speed caused by inappropriate operation of the autopilot.[6] As a result, the passengers experienced a motion, as if they were flying inside a "coffee shaker" on the move (RCAI 2000). The flight was in a severe emergency, and silence was crucial for Ifaistos Limnios to hold a very heavy control stick and land the plane visually. But seven passengers, including the Greek Foreign Affairs Deputy Minister, were killed, and everything inside the cabin was destroyed, as Ifaistos Limnios remembers.

The pilot-in-command was consistent on stating the mechanical warning signal and led in his logbook, regardless of how many times he was repeating himself or was neglected by the manufacturer and partially from his supervising authorities, as Ifaistos Limnios noted emphatically. The flight attendant prompted the "seat belts fastened," but she could not enforce it, as it was the Minister and his son who did not comply, and they could have survived the 24 s of "shaking." In her testimonies she was consistently using the metaphor of the airplane as a reactive *wild horse*.

[4]In the Falcon 900B SX-ECH accident investigation (RCAI 2000) and series of judicial proceedings (Tsolakis 2013, pp. 328–337) several determinants of the situation remain unresolved as the protagonists reveal. But, the "pitch feel" warning signal and its impact in controlling the aircraft and choosing the maximum speed (with the autopilot disengagement after certain speed levels) was a repeated entry in the flight logbook remarks from the pilot-in-command, with no provision from the manufacturing company till this tragic event. The accident investigation team members for this crash were important informants in this study, with long discussion days regarding my topic.

[5]Aircrafts are in constant monitoring for their operating and maintenance record with a service history log which ensures a memory function in their operating history. Eurocontrol's skybrary electronic depository explains "The safety enhancement *SE170: Monitoring of Service History* ensures that maintenance task difficulty data is collected and reported to the original equipment manufacturer (OEM) and proper maintenance is being performed to ensure aircraft systems continue to function as designed. The Federal Aviation Administration (FAA) was tasked in Output 1 of SE-170 with publishing a 'best practices' guide to standardize how these processes should be implemented by operators and OEMs." (data from http://www.skybrary.aero/index.php/SE170:_Monitoring_of_Service_History)

[6]Ifaistos Limnios and other pilots proposed for investigation of the potential of an electronic warfare magnetic field that was activated in the area due to military missions that were conducted in that airspace. In this view the sudden loss of control during decent would be the result of electronic interference; this is a scenario that was difficult to verify, despite some available indications from other flights that landed in Bucharest airport, on that day.

Consequences in the Media and the Industry The Falcon "shaking" VIP flight, as Ifaistos Limnios sometimes calls it, established new standards in the Hellenic Civil Aviation and Hellenic Air Force VIP operations, at an institutional level, for negotiating with silence and speaking up to everyone who needs to hear (regardless the level of statesmanship, industrial authority, or even familiarity and friendship). Moreover, the flight covered media headlines and stories, a series of hearings and trials, like those analyzed by Dekker (2010). Aviation safety is, to a large extent, built upon the trusting pilots, air traffic controllers, and other aviation professionals in the process of accident investigation. There is a growing trend to criminalize these same people following an accident investigation a trend with implications for aviation safety (Michaelides-Mateou and Mateou 2010).

The protagonist, the VIP pilot Ifaistos Limnios who managed to land a plane that was "shaking" prior landing against all the odds, spent 4 days (in Greece and in Jordan) in conversation with me regarding the flight. Ifaistos Limnios was stigmatized from the public opinion's contradiction, although he was not found guilty in court; how could he be responsible for deaths (of people who were not sited with seat belts fastened) while he was capable to land and save the Falcon plane from a crash? He felt trapped in a scapegoating process with the sole responsibility being to activate the "fasten seat belt" sign.

When he learned about a similar incident (with the same technical malfunctions) in the next Falcon model that was grounded with an emergency airworthiness directive on 2011,[7] he felt almost relieved and revitalized professionally, "I told you, they did not do anything, and it happens again" he said. But he was always voicing his opinion to everyone involved: From the Prime Minister to the manufacturing company and the airliner in his own organization. That was his message for the others, as part of his value system, the piece of knowledge he would transfer. That is why despite the severity of the incident, he was able to continue his flight career, unlike what happened with pilots in other accident cases. He had commercial pilot appointments in other European countries and Russia being selected in VIP airliners or business tycoons. Ifaistos Limnios's bottom line was that "when I speak of love, I think about family and my aircraft. When I am holding the controls, the plane unifies with my body; I manage the plane as if I am alone in the universe." That said, Ifaistos Limnios added that when he realizes a potential distress, he explicitly addresses a "standby" to the controller to remain silent and proceeds when he thinks he is done, in case he needs support. In Ifaistos Limnios's case, speaking up was an individual decision of professional ethics, accountability, and integrity (Dekker 2012). And it made a mark in the minds of the aviation professionals which participated in the study for this book.

[7] It was June 17, 2011 when EASA issued an Emergency Airworthiness Directive (AD), prohibiting flight operations of Dassault Falcon 7X jets, due to an incident involved a Falcon 7X which experienced an uncontrolled pitch trim runway during descent. The crew succeeded in recovering a stable situation and performed an uneventful landing (data from Aviation Safety Net at http://news.aviation-safety.net/2011/05/26/easa-grounds-all-dassault-falcon-7x-aircraft-pending-incident-investigation/).

Checking on the Replies Ifaistos Limnios' beliefs were replicated in later conversations I had with a member of the accident investigator team of the HAAISB, namely, *Megas Alexandros*. In the start Megas *Alexandros* had a controversial opinion for Ifaistos Limnios, but soon after the court trials, they became close friends, with mutual respect and appreciation. Both figures are aviators with a strong commitment to "do their job," as duty, as an ideal from their own points of view. Megas Alexandros investigated all flight parameters of the Falcon flight, using also the RCAI (2000) report, and realized that landing the Falcon plane was a *miracle with virtuosity*, but he still thinks that Ifaistos Limnios forgot to activate the "fasten seat belt" sign at an earlier stage before the rapid descent; and he spent almost 5 years dealing with the Falcon flight, due to the series of repeated judicial proceedings and appeals, in the form and intensity analyzed for other flight litigation cases from Michaelides-Mateou and Mateou (2010).

Megas Alexandros echoes the aviation conscience, in several parts of this book, as he is an iconic figure in accident investigation, after a career as fighter pilot in HAF, a career as civil aviation pilot in Olympic Airways airliner, and thousands of flying hours in several different aircrafts. He graduated from Air Force Academies in Greece, USA, and Sweden and specialized in aviation safety in the USA and Sweden. His contributions were valuable inputs to control questions and replies of other respondents in this study.

3.1.2 A Fatal Silent Flight: The Helios 522 "Ghost Plane" Flight

Another paradigmatic flight[8] – the so-called silent ghost plane – laid the foundation of rethinking the concept of silence, in at least two ways (inside the plane and outside the plane). A plane was called from different control ATC authorities with no reply (silence to others), and the lack of open communications between cabin crew members and flight deck deteriorated the situation (silence or sterilization inside). It was a challenge of the sterile cockpit rule and was labeled in the media, as the Greek "9/11" threat.

Hypoxia, Silence, and Renegade Alert Helios Airways Flight 522 in a Boeing 737 originated at 9:07 a.m. from Larnaca, Cyprus, on August 14, 2005, with a history of minor depressurization issues, a contract pilot, and substitute chief purser. Three hours later, the flight crashed onto a mountain in Grammatiko area, outside Athens, Greece. The pilot and copilot seemed to have missed the checklist entry for a crucial air conditioning switch which remained in "Manual" (instead of "Auto")

[8] It is a unique accident in world aviation history. Only two other accidents with similar conditions, in small scale flights, have been reported: Beech Super King Air 200 VH-SKC, September 2000 in Australia (see https://www.atsb.gov.au/media/24344/aair200003771_001.pdf) and Sunjet Aviation, Learjet Model 35, October 25, 1999 (http://www.ntsb.gov/investigations/fulltext/aab0001.html).

position, because of an early ground pressure check before the flight from the airliner mechanics. They set the autopilot to reach cruising altitude towards the Greek Area Control Space,[9] which was the next checkpoint with a ground control station, while approaching the Greek islands in the Aegean Sea. It took just 13 min for the air pressure to drop as the plane was climbing. Oxygen starvation, known as hypoxia, was developing as an insidious process; it impairs human judgment without noticing it. They communicated with false awareness of the situation. As a result, pilot and copilot misinterpreted the true nature of the warning horn that was blaring on the flight. They thought their electronics were overheating, and they became unconscious in minutes. In the main cabin, however, passengers settled down sleepily, and the cabin crew saw oxygen masks sprang down automatically due to the loss of cabin pressure.

The plane was still climbing with the autopilot, instead of carrying out an emergency descent. The cabin crew was unable to walk in to talk to the pilots, as the cockpit entrance was locked.[10] If they had had easy access, they could have provided a clear picture of the passenger cabin, and the captain would have been alerted to use his own oxygen mask (Griffioen 2009). Both Cyprus Air Traffic Control, at Nicosia, and Greek Athens Area Control Center were repeatedly and desperately calling to contact the aircraft. The Greek Air Traffic Control authorities asked with adjacent civil aviation flights to call but still with no reply. But no reply was received nor recorded (HAAIASB 2006). Silence in this case is perceived as suspicious threat (as if it was a deliberate "no response"), and the flight was declared as potential "Renegade flight" (Tsolakis 2013). At 11:24 a.m. that day, two HAF F-16 fighter

[9] On December 7, 1944, the International Civil Aviation Convention was signed in Chicago, providing for the establishment of an International Civil Aviation Organization (ICAO). The ICAO was to delimit the areas of responsibility for air space control for its member-states (Flight Information Region-FIR). The Athens FIR was delimited within the framework of the European Regional Aviation Conferences in 1950, 1952, and 1958. The Athens FIR covers Greek national air space plus some areas of international air space. Turkey was present in the abovementioned regional conferences and accepted delimitation of the air space for which Greece was granted responsibility. In accordance with ICAO rules and international practice, Greece requests that all civil and military aircraft should submit flight plans prior to their entry into the Athens FIR for reasons of safety for civilian flights. The Greek (Athens FIR) was delimited within the framework of the European Regional Aviation Conferences in 1950, 1952, and 1958. The Athens FIR covers Greek national air space plus some areas of international air space.

[10] Following the events of 9/11 aircraft hijacking and attacks, the air transport industry and airlines in particular have instituted a range of physical and procedural defenses to deter and prevent such acts. For larger airplanes, regulatory changes followed. With effect from November 1, 2003, ICAO Annex 6 was amended so that under Chap. 13.2.2 "all passenger-carrying aeroplanes of a maximum certificated takeoff mass in excess of 45,500 kg or with a passenger seating capacity greater than 60 shall be equipped with an approved flight crew compartment door that is designed to resist penetration by small arms fire and grenade shrapnel, and to resist forcible intrusions by unauthorised persons. This door shall be capable of being locked and unlocked from either pilot's station" (from http://www.skybrary.aero/index.php/Flight_Deck_Security). Of course locking the cockpit reduces cockpit-cabin communication and affects in-flight situational awareness of both ends.

aircrafts scrambled and established visual contact with Helios 522 flight.[11] They reported that the first officer was motionless at the controls, the captain's seat was empty, oxygen masks were dangling in the passenger cabin, and there was no movement. At 11:49 a.m., a flight attendant entered the cockpit and sat down in the pilot's seat. Before he had much of a chance to try, after several attempts to enter through the code-sealed cockpit entrance, both engines run out of fuel, and the plane crashed with all 121 passengers and crew. There were no survivors. The scrambled F-16s were reporting, in other words communicating the situation of Helios flight from the outside.[12] Depending on what they report and the knowledge level they establish, the Tactical Chief of HAF, the Chief of HAF, and the Chief of the Joined Chiefs of Staff are in open line with the Defense Minister and the Prime Minister who have the authority to decide whether to shoot down the aircraft or not. It is a decision with no return for all participants in the aviation complex and shows how communication makes it irreversible.

The Agony and Other Consequences Megas Alexandros, the accident investigator mentioned in the Falcon flight of Sect. 3.1.1, shared with me the agony in the mind of the Prime Minister when the flight was still over Athens: "For nothing in this Earth, I would want to be in his position, to have to decide for those people lives if needed. Imagine, also the feeling of the F-16 pilot who was observing the Helios flight unable to help them. Or the feeling in the case that he had to shoot down the plane following orders," Megas Alexandros commented on the severity of the Renegade response protocol. He wanted to emphasize that responsibility for human lives is the ultimate care in aviation safety and contains a moral judgment call which supersedes political or governing power. And communication with accurate observation was the key factor, since the initial impression was that is a real hijacking flight (in several Greek and European media outlets the references was about a "false Greek 9/11" in the end).

[11] A situation where a civil aircraft is used as weapon to perpetrate a terrorist attack is usually referred to as a Renegade (when substantial deviation from the flight plan is identified, or the plane is not following the Area Control Tower instructions, or does not reply in calls and remains silent). The Renegade protocol requires fighter jet scramble for identification and interception. In many cases bi-lateral agreements have been set up to better coordinate cross-border incidents. In most European countries the number of interceptions due to communication loss at least doubled after "9/11". An atypical of those cases, with loss of communication for about 2 h, is the tragic crush of the "Helios Airways Flight HCY522" as investigated by the Greek AAIASB (at http://www.aaiasb.gr/imagies/stories/documents/11_2006_EN.pdf) and is in the mind of all the air traffic controllers that were interviewed for this study.

[12] Five minutes before the crash another dyad of fighter planes from the quick reaction alert squadrons, two Mirage 2000–5s, were ordered for a 5 min readiness to scramble to replace the two F-16s that were running out of fuel due to the maneuver and holding patterns around the Helios jet. The Renegade case of the "ghost plane" was already in the briefings of two major HAF bases, 111 CW for the F-16's and 114CW for the Mirages. The incident became a hot communication topic in the informal mouth-to-mouth network that is resilient in HAF's micro culture, as in most active Air Forces.

But the scrambled F-16 pilots established an additional visual communication channel in which the message was transmitted from the outside using the visual input (and the recordings for more hindsight information, after the crash). Notice, at this point, that without the establishment of that new channel of communication, there would be no message, no information flow and no validation of the situation, and as a result no adequate knowledge for decision-making.

The Helios 522 flight and its story is a landmark case, attracted world media coverage,[13] and prompted several changes and recommendations (to HCAA, to EASA, to Boeing Co, to ICAO, and to NTSB among others), as they derive from the accident investigation (HAAIASB 2006). One of these recommendations was implemented immediately[14]: To apply a software warning flag counter on the radar screens of Air Traffic Controllers of Athens Control Approach Center to be activated when an aircraft is not reporting position more than 3 min after asked. This is the silence-breaker counter, as they now call it, informally. As I have noticed during my visits in the Athens Control Approach Center, it is like a headlight they are carrying during their shifts.

For another time, this tragedy confirms the open channels of communication which operate in the globalized and interconnected aviation industry, not only about an event but also in the follow-up phase of implementing the suggested recommendations. The Helios plane was flying for 2 h with everyone (but one or two) unconscious because of oxygen starvation. Immediately before that, the cabin crew could not communicate with the flight deck and the plane and, as a whole, never responded to repeated incoming communication signals. It remained silent and generated a hunted story for the "ghost plane" in the minds of all participants in this study. When silence prevails in a flight, controllers remember it, fighter pilots in readiness alert tell stories for it, and the government cabinet wishes not to be confronted with the irreversible decision.

Concluding Ifaistos Limnios introduced a landmark case of the "shaking VIP flight," in which silence had a deeper content in organizational reporting: To speak up regardless how many times you are not heard and address the appropriate managerial hierarchy who is responsible for accountability. He reminded the increasing trend of scapegoating for human error which has consequences in aviation safety, as well as the complexity of civil aviation with economic, market, and political concerns.

---

[13] The Discovery Channel, National Geographic, and Near-Miss Productions among others produced documentaries. Also, in the Guardian newspaper in the UK (December 18, 2006), Andrew Weir's position article approached the Helios 522 flight from several interesting dimensions; he pointed out a safety and security problematic that represents a vivid discussion in aviation and society at large regarding cabin-cockpit contact (at http://www.theguardian.com/business/2006/dec/19/theairlineindustry.travel).

[14] A detailed analysis of these recommendations and their impact is featured in the article "The Tragedy of Helios 522 at Grammatiko," January–February 2007 issue of the *Remove Before Flight* specialized magazine published by 11Aviation Publications, Athens, Greece (at http://www.11aviation.com/).

Megas Alexandros described as an outsider the Helios "ghost plane" flight; it maybe one of the handpicked cases in which the hindsight mechanism (there are no survivors and very limited data from the recordings exist) is necessary to communicate the silence of a haunted story. Helios 522 flight redefines what it means to communicate in a timely manner, without waiting silent for feedback. Besides that, it started the discussion for the open channels of communication between the cockpit and the cabin crew members. These two flights constructed a system of reference regarding silence for all responders who were on their duty when the accidents occurred.

3.2 From Theory to Scenarios with Protocol Analysis Lessons

Besides the two landmark flights, the study conducted for this book included two real-flight scenarios suitable for aviation practitioners to comment and disclose their mindset and field expertise. In the process of conducting in-depth interviews to capture the core knowledge of their field, I used several lessons from protocol analysis to facilitate effective responding. In protocol analysis, subjects are trained to think aloud as they solve a problem, and their verbal behavior is analyzed and used to infer the rules being used, as well as various knowledge states concerning the topic (Newell and Simon 1972).

When respondents reply the process went deeper, as I had several interactions with them to elicit their own practices and experiences, their knowledge expertise beyond the content of their training, books, and manuals. The interviews started with an informal phase to speak about their careers, my involvement in the sector and my experience, and their opinion on the importance of communication in their work. The next interview phase was to get a questions/answer interaction to understand what the most appropriate format for these kinds of interviewees was. As a result, a first version of a questionnaire was created for air traffic controllers in training, and the final version with a selection of scenarios was prepared with the review from an experienced accident investigator and fighter pilots who were faculty members in the HAF's Air Force Academy. Informants in this study are not subjects but are considered high-skilled practitioners at the level of experts. This is indeed true particularly for a certain group of pilots-participants who are decorated in international Air Force competitions and drills (e.g., NATO contests), as well as accident investigators with engagement in analyzing hundreds of accidents with global recommendations.

Two Phases of Interaction After the first site visits and exchange of views and iterations regarding "what to ask and in what manner," I finalized a selection of scenarios accompanied with a very small number of personal/opinion questions. The functionality of those shorter questions was to be used as probing devices for respondents, to complement one response with another in related instances, and to trace potential inconsistencies when they respond from the point of view of their role (controllers were asked for the position as if they were pilots and pilots as if they were controllers or copilots or pilot instructors). In the closing iteration, regarding the configuration of a questioning path for the carrying out the interviews, I was able to cultivate and instruct responders to verbalize their thoughts in the way that they complete their tasks in the certain moment of time when, for example, pilots or controllers are confronted with a hesitation, a repeated entry in an operating error log, the reluctance of an authoritative passenger to follow purser's suggestions, or a developing emergency when the controller is standing by after the initial declaration of a situation of a crew member, to name a few of the tasks.

The interaction with responders followed two phases: In the first one, in a certain scenario, we had to discuss the actions of the pilot or controller as they are performed in a prescribed algorithm (a specific set of instructions for carrying out a procedure or solving a problem with a start and an end), and in the second step, we had to connect the actions taken, as functions that show concepts of the operating environment, with communication interactions. Sometimes in this process of their "reply and discuss," responders did not offer a one-to-one correspondence of their verbal report with the processing steps, since the skills they possess and level of expertise make them take *their own knowledge for granted when they respond*, so their attention may be short-circuited with their acquired skill set. That is why I have conducted a couple of follow-up discussions with ten of the keenest responders to get their rumination (a week) after the completion of the scenarios; in Chaps. 5 and 6, data analysis has additional probing questions and notes on background or regulatory tips and associations useful to unpack such occurrences.

A relevant example of using some of the tips of protocol analysis is the detailed description of a fighter pilot who did not speak up on a navigational decision about approaching an airport without using local ground approach positional systems; he had to explain the operating principles of a certain positioning device (viz., TACAN), in order to connect it with his reaction of no-talk in a two-seat aircraft training flight with an instructor pilot. While he was thinking aloud about a no-talk instance of his, his recollection became an algorithmic reenactment of the flight; we had to go through the list of his tasks, together, since the mission briefing phase before takeoff, so that his thoughts mediating the completion of the task did not distort the story he was able to represent.

He related this flight with the scenario of "not speaking up" despite his belief that a small gap in the routine flight planning phase generated a doubt, and the instructor pilot (who is in command) was proven to have the same uncertainty. For this young second-seat pilot, the mission connected with silence when his superior officer (and pilot in command) spoke up in reply to his urging question. After an initial touch on the ground (landing with no stop) and before continuing the next "touch and go's" (immediate takeoff with the same rolling in the runway). The two pilots realized a mutual silence period (the one not asking, the other not telling) which evolved to a dependable mentoring relation after their interaction. The responder's thoughts mediating the completion of a task are used to reconstruct a reply for the situation which was unfolding during the actual flight and can, therefore, be accepted as valid data on the responder's thinking of what he was doing.

Lessons from Protocol Analysis The central assumption of protocol analysis is the instruction to the responders to verbalize their actions in a manner that doesn't alter the sequence of thoughts mediating the completion of a task and can therefore be accepted as valid data on their thought process (Ericsson 2002). The method involves making a detailed record of a person's verbal report while they are engaged in carrying out a task, for example, doing a mental calculation, solving a problem, making a decision, or interacting with an automated device. In this study, the reply was a retrospective representation of the pilot's tasks as if he was completing those tasks in front of the scenario with the framing context (of what flight or incident of his practice is closer to the scenario description) he determined to reply to the scenario and the interpretation he selected for his silence in interaction. He was the one who practiced silence and decided to break it after a first landing approach judging that his uncertainty was a safety risk more important than the reservation he had for the status of his instructor. Unlike that, other responders (like controllers) received silence and replied with another point of view in the same scenario. Hence, these lessons of protocol analysis assumptions were prompting devices for a more relevant reply in the scenario.

The verbal reports are then written in plain language, segmented into statements, and coded (using episodes, based on participants' working on a particular subtask) in reference to the concepts which relate to the core inquiry of human communication in this study (as in Fig. 4.2). This phase connects with the coding scheme of a GT approach as this study emphasizes "what is communicated" and limits the search of the "underlying cognitive processes" which would be the aim of a pure protocol analysis study. Participants here describe and explain "what they did and not what they ought to do or (just) were taught to do." That is why in several cases responders deposit their own deviations and the way the correct errors that accepted to have made. As it happens to a GT approach, also protocol analysis principles, note that a subjective interpretation of verbal reports must be recognized. Since the responders are not in an experimental control environment, the use of a small group of selected experts as evaluators and the deep closed reading of their data are deliberate choices I made to mitigate that level of subjectivity.

In the following section, two examples of main scenarios from the third-stage questioning tool are presented as the basis of final data and concept collection of a GT approach. The full document with the scenarios can be found in Appendix A.

3.2.1 The Scenario of Silence

The first main scenario discusses silence, using research findings from the practice of silence in an airline organizational setting. The informants-interviewees who provide the raw data for this book, after a certain discussion period, offer their commentary in writing. The point of view of different actors (pilots, controllers, crew members, accident investigators) determines various manifestations of silence and events in which silence was decisive of the conclusion of the flight or the safety of passengers on board.

Since the early years of (mainly) civil aviation, crew members' reluctance to speak up and the resultant silence have posed a serious threat to flight safety. Silence, in the mid of the responders, may be observed inside the airplane cabin (with passengers, pursers, and flight attendants), inside the airplane cockpit, and between cockpits (in crisis situations) or may be measured institutionally with the flight log reports where the crew has to fill any noteworthy information (for the airlines and the manufacturer, such as mechanical problems, operating environment problems, technical inconsistencies and so on). Also, responders are asked to take the point of view of their role, so that silence would emerge as a phenomenon with multiple dimensions: Silence from whom, addressed to whom, used by the controller or the crew member, and perceived by the controller, crew member, an external pilot, or the airline operator. Since a flight is accomplished in a collaborative environment, NASA introduced a managerial requirement in 1979, promoting the cockpit resource management that evolved to crew resource management (CRM). Symbolically, the prompting device of CRM was the triplet: communication, communication, and communication.

The scenario listed below, in the form of research report, investigates the theme of silence as introduced by Bienefeld and Grote (2012) but extending it further:

The Main Scenario of Silence

Several accidents have shown that crew members' failure to speak up can have devastating consequences. Despite decades of crew resource management (CRM) training, this problem persists and still poses a risk to flight safety.

To resolve this issue, we need to better understand why crew members choose silence over speaking up. We explored past speaking up behavior and the reasons for silence in 1,751 crew members from a European airline, who reported to have remained silent in half of all speaking up episodes they had experienced.

Silence was highest for first officers and pursers, followed by flight attendants, and lowest for captains.

In this scenario, the focus is on silence as a contributing or causal factor of aviation accidents (crises) and on answering the "why" question, as if silence and speaking up are the two sides of a coin; those crew members could speak up but decided not to, in half of the times that had the opportunity or the need. It seems that the most extrovert members (flight attendants/pursers and captains) had a higher rate to speak up in cabin crew communication. The commentary and feedback regarding this scenario are unravelling in Chap. 5, in which selective respondents share their pieces of knowledge.

3.2.2 *The Scenario of Hesitation*

A second major scenario used has to do with the experience of hesitation in interaction that may leave incomplete information, truncated messages, generate uncertainty, and result to insufficient knowledge in the cockpit-controller interaction.

The scenario is taken from an anonymous report of a real (pre-) flight situation in NASA's Aviation Safety Reporting System (ASRS 2013). It develops in three parts: (a) A captain (as pilot flying or pilot in command in civil aviation) reports what he heard from the controller and the resulting dilemma and what he did in a rolling movement to takeoff; (b) after the action, the captain is reporting and explaining the position taken, choice made, and action performed (so he is meta-communicating on the interaction); and (c) the first officer (as a pilot not flying the airplane) is commenting and evaluating on the incident and the action of the pilot flying the aircraft. This is also another level of meta-communication (closer to debriefing) for the initial interaction and the incident. It is an engaging scenario for the responders with the triggering question "what would you do?"

The vocal but (still) nonverbal dimension of speech channel of communication has to do with how one says something and not with what one says. Among the voice qualities, prior to actual talking, the language, pause, and hesitation are perceived and understood in the broader verbal interactions. Adjustment is necessary for both sender and receiver ends to communicate. That is why a second scenario from an actual reported flight situation is introduced, to investigate the transition from silence to language, via paralanguage. In this scenario we move from clear nonverbal instances to hesitation (which has a vocal component) as an indication of bypassing in the cockpit-controller interaction or as a contributing factor of aviation accidents. Why pilots or air traffic controllers hesitate in their communication flow? How do both deal with hesitation to avoid miscommunication towards an unfolding incident, with or without the rules? Let us see how the scenario is laid out with the captain of the anonymous report, his actions, and the reaction of the first officer.

Hesitation Scenario Part (a)

Captain reports: On takeoff roll approaching 80 knots, the Tower Controller called us and said in a very slow, unsure voice, "[Callsign 1…2…3…4…] (pause)." He sounded as if he had something to tell us but did not know what to say.

We both noted a tone of concern and hesitation in his voice as if he was still unsure of something at that moment. We were lightweight and had 13,000 feet of runway ahead of us. We had to make an immediate decision.

The civil aviation pilot (the captain) reports that he has ended the taxiway movement, continued to the runway to take off, and started rolling to develop the necessary speed. The call from the air traffic controller was made with a pause and murmuring the call sign (which identifies the aircraft in their interactions; it is usually the carrier name with a number). Although the message was incomplete, the tone was thought-provoking, and timing was critical as they have started the speeding up for the takeoff process. They still had some route to go, and they had to decide with no other input, just their airplane data (fuel to avoid fire risk and the remaining distance to be safe). What did the Captain do?

Hesitation Scenario Part (b)

The reporter's action: I elected to initiate rejected takeoff procedures.

During deceleration the Tower Controller said, "Disregard."

The sound of one's voice, the tone and force, all convey a message. I did not like the message.

I was receiving and could not gamble that he was trying, but unable, to warn us of something ahead.

would take the same action again.

The captain decided to take the incomplete call as a reason to reject the takeoff and started the deceleration in order to stop the aircraft (due to its volume, weight, speed, and technical configuration the airplane cannot stop instantly with breaks). At that point, the call from the controller continued with a "disregard," as it was nothing to talk about, to ignore the call. The captain is explicitly saying that uncertainty was something he could not evaluate, in case it was a warning for a potential problematic situation ahead (they were cleared to take off and the call would project a reason to revoke that clearance). In the original callback report, the first officer in the same airplane posts their position, as an insider who was able to receive the same incomplete call. What is their evaluation?

Hesitation Scenario Part (c)

The first officer's report on the same incident: I believe the rejected takeoff was the right thing to do.

When you get a call from Tower at that point in the takeoff roll, the first thing that pops into your mind is "something's wrong."

In the few seconds before he finished his thought, we were left to guess what the call was about.

We were still relatively slow speed on the roll, so the captain did what was prudent and safe by rejecting.

The first officer is fully confident for the action of the captain, although there is no information whether they have discussed it or not. It seems that, since the flight was actually accelerating to take off, a call from the tower breaks the sequence which seems to be uneventful. The potential of anything unexpected is a call for being prudent than being sorry. The hesitation scenario explained above is the discussion topic for the discussants who shared their experience for this study. The inquiry was to evaluate the hesitation issue, to explain their action as if they were pilots, and to suggest ways to repair hesitation and other interference for the first officer. The commentary and feedback regarding this scenario is unravelling in Chap. 6, in which selective respondents share their pieces of knowledge.

The reader should have in mind that a successful (or accomplished) flight is a systemic result of the interactional accomplishment of human performance. This book is approaching knowledge construction following a communication orientation via the basic model of communication (Shannon and Weaver 1949) and notions from the helical model of communication (Dance 1967). Every interaction is based on a first event, time is important, and communication process is a product of what actors have learned inward in ways that permit growth, discovery, and cooperation towards a shared meaning. The nature of flight communication which considers meaningful information as the specific identification of a signal from the volume of "noise" extends the basic model, towards the interhuman communication network (Leydesdorff 2016). While exploring how experts develop and apply their knowledge skills, the use of the two paradigmatic flights and the two real-flight scenarios generate an instrument to trace their steps.

3.3 Human Factors Framing Concepts in Flights

The landmark flights and scenarios described in Sects. 3.1 and 3.2 easily provide specific knowledge elements when considered through the lenses of human factors in the related communication process (in the framework of Sect. 2.3). Human factors goal is to understand the interactions among humans and other elements of a system and apply theory, principles, data, and other methods to design in order to optimize human well-being and overall system performance. So, human factors deals with the physical, cognitive, social, organizational, environmental, and other relevant factors involved with the creation of things and processes that effectively support people in their work and living activities. The flights and scenarios discussed in the chapters of this book are analyzed with the experience of expert aviation informants who have in mind such a systemic view and operate with special concern in the principles of human factors.

Human actors' behavior and reflections as part of human factors formulate a thorough understanding of machine interaction and human conduct based in cognate processes (Cicourel 1973; Kendon 1977), interaction rituals, and internalization of conversation in group work and leisure (Goffman 1967; Collins 2005).

Along these lines, organizational human factors include topics of communication, crew resource management, design of work, design of working times, teamwork, participatory design, community ergonomics, cooperative work, new work paradigms, virtual organizations, telework, and quality management. All these aspects realize in the context of an organization, like airline companies, air force services, aviation agencies, and institutions. Organizations are structured groups of people, working together, using technological artifacts and procedures to achieve compatible goals (Devito 2011), and utilizing procedures and above all communication that ensures information flow (Barnard 1938/2005). In this study, silence communication and nonverbal aspects of communication enrich the human factors literature with concrete knowledge of the aviation situation.

Human factors, as a discipline in state organizations and public administration, has a wide scope as most of employment is found in those organizations, their complexity increases, and they exert influence over public policy. A variety of topics on critical environments have used human factors: vehicle collision avoidance (Polychronopoulos et al. 2007), patient satisfaction in the health sector (Moss et al. 2002; Rowland-Morin and Carroll 1990), nuclear safety (Moray and Huey 1988), etc. In the Air Traffic Management (ATM) system of pilots and controllers, human factors described the evolution in complexity and flexibility of workload environment (Laios and Giannacourou 1995). In the current study, the association of situation awareness (as explained in Sect. 4.3.2) with communication realities contributes to extend the human factors literature.

The International Civil Aviation Organization (ICAO 2011a, b) has also taken a human factors approach by emphasizing the role of the people who have become instrumental in aviation safety as "live ware." In the current book, the term includes the human actors in the cockpit and stakeholders beyond that cockpit towards the air traffic controller (as in Edwards 1988). The European Air Traffic Management system (EATM), known as Eurocontrol, issued its guidelines for best practices in 2007. When introducing human factors the guidelines include the user-actor (human in system) in the cockpit and ATC working environment (as context), organization processes and staffing (supporting users), procedures, roles, and responsibilities (SOPs, flight instruction terminology, communication manuals) acquired through training and development of team building via concrete communication adjustments, as relationship cultivation through communication.

This is a complete "systems view" of the working environment with the human actor in teamwork. Teamwork is accomplished via communication, applying procedures – roles – and responsibilities cultivated in training phases that develop personnel necessary to populate and operate an organization around the clock. When shifts change, appropriate transition and vigilance is necessary, so people conduct their roles and tasks although actual individuals change. The human factors field in aviation is also considered in driving a "systems level" with the following six areas of concern (Eurocontrol 2008): human-machine interaction (HMI); organization and staffing, training and development; procedures, roles, and responsibilities; teams and communication; and recovery from failures. HMI is by itself a form of communication that is mediated by human actors in layers in the cockpit and in ATC.

Trust with Humans and Machines In this book, the term "human actors" is more inclusive to describe interlocutors that are physically present inside the cockpit (pilot and copilots), as well as external players mainly air traffic controllers (ATC), crew members, airline operators, and possibly pilots of other flights. All these layers cannot accomplish their tasks without establishing and relying to trust. There is a broad literature on trust in automation, where most agree that accuracy of information is crucial in building and retaining trust. Accuracy is a knowledge attribute that enforces trust. It is an aspect of collective self, described as reflex (Salem 2009); a fixed expectation of culture in the flight mission, sometimes as being learned and situational. Trust is also considered as a resource of social capital, authority, and social support. As Salem states it is an "interaction variable leading to a probability to engage [with] another."

Expected benefits from trust include better task performance (Golembiewski and McConkie 1975) and the ability to cooperate (Deutsch 1958, 1962; Argyle 1991). There are a number of taxonomies for trust in human-machine communication and interaction (Barber 1983; Rempel et al. 1985; Muir, 1987, 1994; Muir and Moray 1996). These taxonomies tend to distinguish three categories of trust (Lee and Moray 1994): (i) trust based on observed consistency of behavior (persistence or predictability), (ii) trust based on a belief in competence (competence or dependability), and (iii) trust based on faith in purpose or obligation (confident responsibility or faith).

While faith-based trust is difficult to establish for computers and automation, these taxonomies suggest that trust by consistency and competence can be achieved by explaining the purpose, capability, and reliability of the software or the system to the user. In this book, trust is taken for granted in the interaction between pilots and air traffic controllers but needs to be sustained during interaction when situation awareness evolves. Moreover, in the cockpit acts of conversation, the issue of trust in automation and reliance to each other introduces another dimension (Lee and See 2004; Ostrom 1998; Muller 1996), consistent with individual member tacit knowledge (Polanyi 1966). Thus, when human actors operate mechanical and communication devices in flights, they have to build trust in a mutual manner with all the other participants, in each stage of a flight mission.

References

Argyle, M. (1991). *Cooperation: The basis of sociability*. London: Routledge.

ASRS (Aviation Safety Reporting System). (2013, January). *What would you have done? Callback 396 (situation #4). NASA Aviation Safety Reporting System*. Available at http://asrs.arc.nasa.gov/publications/callback/cb_396.html

Barber, B. (1983). *The logic and limits of trust*. New Brunswick: Rutgers University Press.

Barnard, C. (1938/2005). *The functions of the executive* (30th anniversary ed.). Cambridge MA: Harvard University Press.

Bienefeld, N., & Grote, G. (2012). Silence that may kill when aircrew members don't speak up and why. *Aviation Psychology and Applied Human Factors, 2*(1), 1–10.

Cicourel, A. (1973). *Cognitive sociology: Language and meaning in social interaction*. Harmodsworth: Penguin.

Collins, R. (2005). *Interaction ritual chains*. Princeton: Princeton University Press.

Dance, F. E. X. (1967). A helical model of communication. In F. E. X. Dance, (ed). *Human Communication Theory*. New York : Holt, Rinehart and Winston.

Dekker, S. W. A. (2010). Pilots, controllers and mechanics on trial: Cases, concerns and countermeasures. *International Journal of Applied Aviation Studies, 10*(1), 31–50.

Dekker, S. W. A. (2012). *Just culture: Balancing safety and accountability* (2nd ed.). Farnham: Ashgate.

Deutsch, M. (1958). Trust and suspicion. *Journal of Conflict Resolution, 2*, 265–279.

Deutsch, M. (1962). Cooperation and trust: Some theoretical notes. In M. R. Jones (Ed.), *Nebraska symposium on motivation* (pp. 275–319). Lincoln: University of Nebraska Press.

DeVito, J. A. (2011). *Human communication the basic course* (12th ed.). Boston: Allyn and Bacon.

Edwards, E. (1988). Introductory overview. In E. Wiener & D. Nagel (Eds.), *Human factors in aviation* (pp. 3–25). San Diego: Academic.

Ericsson, K. A. (2002). Toward a procedure for eliciting verbal expression of nonverbal experience without reactivity: Interpreting the verbal overshadowing effect within the theoretical framework for protocol analysis. *Applied Cognitive Psychology, 16*, 981–987.

EUROCONTROL. (2008). What is 'human factors' all about? Eurocontrol white paper. Online. Available http://www.eurocontrol.int/eec/public/standard_page/human_factors_what_is_it.html. Accessed 3 Nov 2011.

Goffman, E. (1967). *Interaction ritual: Essays on face-to-face behaviour*. New York: Doubleday & Co. Anchor Books. (Harmondsworth: Penguin, 1972).

Golembiewski, R. T., & McConkie, M. (1975). The centrality of interpersonal trust in group processes. In G. L. Cooper (Ed.), *Theories of group processes* (pp. 131–185). London: John Wiley & Sons.

Griffioen, H. (2009). *Air crash investigations: The crash of helios airways flight 522*. (Lulu.com Edition).

Hellenic Air Accident Investigation and Security Board (HAAISB). (2006). *Helios airways flight HCY522, boeing 737-31S at Grammatiko Hellas on 14 August 2005*. Athens: Ministry of Transportation & Communications.

(ICAO) International Civil Aviation Organization. (2011a). *State of global aviation safety – 2011*. Montreal: ICAO Publication. Online. Available http://skybrary.aero/bookshelf/books/1723.pdf. Accessed 3 Nov 2011.

(ICAO) International Civil Aviation Organization. (2011b). *Strategic objectives: Safety*. Online. Available http://www.icao.int/safety/Pages/default.asp. Accessed 3 Nov 2011.

Kendon, A. (1977). *Studies in the behavior of social interaction studies in the behavior of face-to-face interaction* (pp. viii + 260). Lisse: Peter De Ridder Press (Vol. 6, Studies in Semiotics. Research Center for Language and Semiotic Studies, Indiana University, Bloomington, IN).

Laios, L., & Giannacourou, M. (1995). Human factors issues of advanced ATC systems. In N. McDonald, N. Johnston, & R. Fuller (Eds.), *Applications of psychology to the aviation system. Proceedings of the 21st Conference of the European Association for Aviation Psychology (EAAP)* (Vol. 1, Chap. 17).

Lee, J. P., & Moray, N. (1994). Trust, self-confidence and operators' adaptation to automation. *International Journal of Human-Computer Studies, 40*, 153–184.

Lee, J. D., & See, K. A. (2004). Trust in Automation: Designing for appropriate reliance. *Human Factors, 46*(1, Spring), 50–80.

Leydesdorff, L. (2016).Information, meaning, and intellectual organization in networks of inter-human communication. In C. R. Sugimoto (Ed.), *Theories of informetrics: A Festschrift in honor of blaise cronin* (forthcoming).

Michaelides-Mateou, S., & Mateou, A. (2010). *Flying in the face of criminalization; the safety implications of prosecuting aviation professionals for accidents*. London: Ashgate.

Moray, N. P., & Huey, B. M. (Eds.). (1988). *Human factors research and nuclear safety*. Washington, DC: National Research Council, National Academic Press.

Moss, J., Xiao, Y., & Zubaidah, S. (2002). The operating room charge nurse: Coordinator and communicator. *Journal of the American Medical Informatics Association, 9*(6), S70–S74.

Muir, B. M. (1987). Trust between humans and machines; the design of decision aids. *International Journal of Man-Machine Studies, 27*, 527–539.

Muir, B. M. (1994). Trust in automation: Part I. Theoretical issues in the study of trust and human intervention in automated systems. *Ergonomics, 37*(11), 1905–1922.

Muir, B. M., & Moray, N. (1996). Trust in automation. Part II. Experimental studies of trust and human intervention in a process control simulation. *Ergonomics, 39*(3), 429–460.

Muller, G. (1996). Secure communication – Trust in technology or trust with technology? *Interdisciplinary Science Reviews, 21*, 336–347.

Newell, A., & Simon, H. A. (1972). *Human problem solving*. Englewood: Prentice-Hall.

Ostrom, E. (1998). A behavioral approach to the rational choice theory of collective action. *American Political Science Review, 92*, 1–22.

Polanyi, M. (1966). *The tacit dimension*. London: Routledge & Kegan Paul.

Polychronopoulos, A., Tsogas, M., Amditis, A., & Andreone, L. (2007). Sensor fusion for predicting vehicles' path for collision avoidance systems. *IEEE Transactions on Intelligent Transportation Systems, 8*(3), 549–562.

Rempel, J. K., Holmes, J. G., & Zanna, M. (1985). Trust in close relationships. *Journal of Personality and Social Psychology, 49*(1), 95–112.

Romanian Civil Aviation Inspectorate (RCAI). (2000). *Final Report on the accident of the Falcon 900B SX-ECH- 14 September 1999*. Bucharest.

Rowland-Morin, P. A., & Carroll, J. G. (1990). Verbal communication skills and patient satisfaction: A study of doctor-patient interviews. *Evaluation & the Health Professions, 13*(2), 168–185.

Salem, J. P. (2009). *The complexity of human communication*. Cresskill: Hampton Press.

Shannon, C., & Weaver, W. (1949). *A mathematical model of communication*. Urbana: University of Illinois Press.

Tsolakis, A. D. (2013). *Dare! A way of life (Tolma! Enas Tropos Zois)*. Athens: 11Aviation Publications.

Chapter 4
Eliciting Expertise, Harvesting, and Representing Knowledge

4.1 Overview of an Applied Grounded Theory Approach

This book is based on the study of human communication as a direct contributing factor to establish the knowledge level necessary to avoid consequences to aviation safety. I use a bottom-up approach, which moves from qualitative data gathering further, to identify relations between communicative behaviors (actions, reactions, and interactions) with iteration and refinement and allow core concepts to emerge with theoretical sensitivity; core, here, means that the concepts have an impact to human communication and consequences for aviation safety. This approach follows the grounded theory paradigm dealing with aviation from an organizational point of view. "Grounded" means that it is following a strategy of bottom-up theory building to investigate the discourse and knowledge constructed (Glaser and Strauss 1967).

I use a GT approach with selective parts from the two dominating streams of literature in the sector, the Glaser's stream and the Strauss's stream, as they evolved in coding (Charmaz 2009; Grover et al. 2014; Jantunen and Gause 2014), theory construction and comparisons (Corbin and Strauss 1990; Glaser 1992; Christiansen 2011; Kelle 2005; Ritchie and Lewis 2004), evaluation (Wagner et al. 2010; Suddaby 2006; Robson 2002; Guba and Lincoln 1994), and iterative sorting and analysis (Muller 2014; Muller and Kogan 2010; Reichertz 2007; Locke 2001). I am putting my interpretation in the real-flight raw data and the scenarios, adding multiple relevant resources, to engage in conversation with knowledgeable experts (Patton 2002). I lean to the hybrid practice of using what works better for the data and the participants of the study. As Kelle (2005) concludes, I tend to agree towards the position that the two schools complement each other. Thus, the phrasing I use refers to "a GT approach or approaches" to elicit knowledge.

T. Katerinakis, *The Social Construction of Knowledge in Mission-Critical Environments*, Innovation, Technology, and Knowledge Management,
https://doi.org/10.1007/978-3-319-91014-7_4

4.2 Application of a GT Approach for Selecting Sources

In the beginning, I explicitly take into account my own position as Suddaby suggests (2006). My personal preconceptions, biases, and view on aviation serve as the main preparation of the study and are explained in the following sections. A process of constant comparison starts with the first level of reading and identifying communication phenomena with human actors. Analysis of data proceeds in search of theoretical categories that the previous phenomena relate to and the decision for new sampling (selection) for types of data tightly connected with the theoretical concepts of communication interaction is taken; it is about what type of data are to be collected next to support the theory construction of how human interaction relates to a communication concepts that establish awareness and promote knowledge with impact to aviation safety.

When exploring aviation, as a workplace, an institutional environment, a communication context, and a social world, I use a variety of empirical tools to acquire insights on routine and problematic moments and meanings. The multifaceted reality of actual flight operations requires multiple sources to discover an emerging theory which matches the data and would work in the real world of aviation. A GT approach applied as qualitative research analyzes words, language, and their implied meanings for the knowledge they contain (Miles and Huberman 1994). In such an approach, extended empirical data are treated as texts with multiple meanings. Data collection acquires theoretical significance with the selection of additional data (events, testimonies, visual productions, activities, experts, etc.) to exposit all properties of the developing conceptual categories and develop them further, at both the individual and social levels. I will take us through each different data source, in a process where the more useful concepts are sustained and others, less compatible with the evolving theory of knowledge, do not continue further. These resources for data gathering, categorizing, and analyzing include the following:

- Personal experiences and preconceptions
- Ethnographic field visits to airports, air bases, control towers, and aviation authorities
- Case studies of landmark flights
- Interviews with aviation professionals
- Life stories in biographical books and oral narratives
- Artifacts, cultural texts, and productions via a special drama series on fighter pilots
- Observational, historical, interactional, and visual texts, in documentaries and media interviews
- Protocol analysis lessons to facilitate the acquisition of insights
- Discourse analysis practices with close reading for the theory construction, sorting, and comparisons

The above-listed resources forward the investigation and emergence of concepts that relate with safety risks and are observed in real-flight situations. Unlike other investigations of human interaction, aviation communication pays attention to each sole flight, for instance, when a single flight is considered as a unit of analysis. Flight operating manuals describe an ideal flight situation, whereas the real-flight operating environment is quite more interactive. Cockpit crew has to interact with a range of human actors on the ground and in the air. These actors send critical information to the crew, expect information from the crew, and affect the timing and configuration of crew's tasks with their demands (Loukopoulos et al. 2009).

My personal experience originates from a long-standing relation with aviation since my high school years, the affinity that society in Greece has with the Air Force, and a strong affiliation with the Air Force, due to my military duty in HAF and USN bases (as explained in Sect. 1.2). This experience has generated a network of relations and few preconceptions about roles, attributes, and practices of the members of the aviation community. Therefore, the process of acquiring and selecting information was accompanied from introspections in challenging my Air Force experience.

A main source of substantive information was the fieldwork in on-site visits, at a conceptual level with the main concerns of the industry and Air Force trends and at the level of empirical data with flight transcripts, regulatory documents, and note-taking in memos which became quite useful during the coding-sorting iterations of a GT approach. During these visits I spent substantive time observing the work of air traffic controllers on airport towers and control approach center, as well as monitoring pilot briefing and mission design while taking notes for my memos. In one of these visits, I was prompted to the transcripts and accident investigation reports of the two landmark flights related to aviation accidents. The use of these two landmark flights as case studies (Helios 522 "ghost" plane and Falcon VIP "wild horse" plane, both presented in Sects. 3.1 and 3.2) augments the theoretical sensitivity of the study with immersion in the data, as a Glaser's GT approach requires. These insights contribute to the knowledge acquired regarding silence and voice communication channels.

Comparisons between segments in this data and expert comments add concepts (like silence, accountability, and nonverbal communication) and relations of communication phenomena with aviation safety, the main topic of this book, as a GT approach closer to Strauss suggests. An entry was repeatedly stated in the flight incident log but was not arranged with fatal consequences, and a persistent no-reply call due to which fighter jets engaged to communicate cockpit-to-cockpit are, both, typical episodes from these two flights.

I was able to conduct a certain number of "compare-and-question" activities via three stages of interviews with air traffic controllers, civil aviation pilots, military pilots, and the comments of accident investigators. Few short question/answer sessions showed the inefficiency of that form of data collection, as open discussions were going off topic (an example of an interaction is shown in Appendix C). In a second stage, an Ishikawa (fishbone) questioning tool (shown in the Appendix B) with another core theoretical concept was used (Ishikawa 1968). The topic was the practitioners' ability to deviate from rules using tacit knowledge. Fighter pilots were inclined to this format, as it was close to the tactical design charts they use but controllers and commercial pilots could not connect with (USAF n.d.; DoN 1993).

Hence, I went to the third and final stage, with scenario-oriented interviews (with limited direct single questions) which became the most applicable tool across different aviation professionals. During the first eight interviews, in the first two stages, the discussion incorporated a personal tone with contexts outside aviation and more related to personality traits. When I switched to the more focused and explicit Ishikawa questioning route, the proposition regarding "knowledge expertise and deviance from the letter of the rule" was applicable in the flexible mission design of fighter pilots, but was not comfortable for controllers and commercial pilots. A communication shift in the clear pattern emerged: Respondents follow rules but also take own initiatives in specific situations as they interpret information of what is around them. Safety is their priority, but they need to communicate their choice of action to the other actors who may not expect that action. The three-stage process was a valuable iteration from data to responders, to forms, and to concepts.

As noted the third stage tool contains scenarios labeled with actual core concepts, as Glaser's GT approach recommends. On the other hand, the scenario structure is a skeleton compatible to the axial coding that Strauss's GT approach supports. Strauss's coding paradigm, with conditions prompting interactions (of pilot and controllers), as well as underlying strategies (used by the actors) with certain consequences, had a full application in the scenario structure. In the following parts of this chapter, specific examples from those stages are presented, whereas the full version is posted in Appendix A.

Secondary but enriching sources were the life stories of pilots who have survived near-crashes and saved the aircraft, as well as stories from fighter pilots who were engaged in (almost daily) dogfights over the Aegean Sea. The first part consists of severe instances like the Flight "US1549" on the Hudson River of New York (NTSB 2010; Kochan 2009) or the Falcon "VIP flight" (Konstantaras 2007; Tsolakis 2013). Crew coordination, pilot initiative and efficient judgment, and controller's limitations were apparent in the civil aviation flights. The second part refers to the specifics of bursting interactions during dogfights which remain in orality (stored in flight recorder and cockpit videos), due to the confidential nature of the missions/channels. I was able to access two examples through my personal security clearance. Despite their orality these narratives disclose crucial skills in aviation communication, like multichannel listening, mother tongue laconic explicit interaction, implementation of rules of engagement via communication, etc. Therefore, such oral narratives have the historicity and materiality of documented information (Turner 2010).

These stories add on the multiplicity of viewpoints that GT approaches encourage in the exploratory nature of this study towards a generalized causal model in the form of "what contributed" to pilots' efficacy in navigating during major accidents and save the planes. How did they communicate their critical situation and know what to do for a safe outcome?

Another piece of data was the drama series collection "Silence in the Air" which I decided to watch, as it was the only drama series for fighter pilots' life in Europe, in broadcast for 3 years with their artifacts and cultural texts (Anosis SA 2000). The series used cultural attributes of the Greek society (like the striking affinity of

HAF in the Greek society, the duty obligation, and ethical practices in interpersonal relations) to project an idealistic image of squadrons of fighter pilots and accentuate the bonding, the mutuality, and camaraderie in their subculture of an air base. After the first 40 episodes, the plot moved to display a contrast between the intense but organized professional service life of fighter pilots and the instabilities they must deal with in their private life.

The professional-personal life balance was the script in several memo notes of mine, while probing in conversations and informed relations of sociality and directness in phraseology and communication, as well as the "rushing-to-return" phenomenon which was present in two scenarios. Note-taking in memos and connecting interpersonal communication concepts like keeping face and emotional stability traits in the workplace were part of a taxonomic process for the attributes collected in initial interviews. Memos and sorting functions are required in iterative mode in both Glaser's and Strauss's GT approaches.

Lastly, observing historical documentaries and media interviews – from stations like C-Span, National Geographic and the History Channel, as well as investigative reporting episodes of the Greek television with former Air Force officers or air traffic controllers and broadcasts of judicial proceedings after accident investigations – allowed me to formulate a more holistic view of aviation, its actors, and practices beyond my own prejudices and frame the aviation safety concept in a concrete way. In Chaps. 5 and 6, there are several broadcasted analogies mentioned in different sections.

4.3 Towards the Core Concept to "Know about"

In this book concepts originate from theoretical notions of social sciences, like the concept of "role expectations," in which participants react according to the expectations of their roles. In this exploratory study, such a concept applies in *ideal-flight* situations where information might compare with phenomena of *real-flight* situations. The concepts of silence and voice are categories observed in real-flight situations and have an impact in aviation safety. A second type of categories relates to specific local knowledge of the field of study of aviation. The various types of signals and the emergence of the perception of situation awareness fall in this category.

GT approaches recommend diverse sources of evidence, in the same way that the material for this book is expanded to compare and analyze a rich depository of data (Wagner et al. 2010). The desired attribute is transferability in the form of expanding theories of knowledge that should provide "a general guide to multi-conditional, ever-changing daily situations" (Glaser and Strauss 1967). The decision about the amount of different types of data collected is evaluated towards the end of the study from the conceptual relations which the sources associate with aviation safety (the main knowledge parameter in this book). By the time when more attributes connect with aviation safety and are associated with communication concepts, then a closure of search for the core component of human communication is reached. The degree to which different types of data converge towards communication attributes determines the theoretical sensitivity trait of GT approaches for both streams of literature (Glaser and Strauss 1967).

4.3.1 Core Knowledge Concept: Aviation Safety and Dimensionalization

The "core category" in grounded theory derives from a constant comparative method of every part of data (codes, categories, properties, and dimensions or different parts of the data). Comparisons with all other parts of the data explore variations, similarities, and differences in data (Hallberg 2006). The literature of GT approaches uses the terms "category," "label," and "concept" interchangeably (Martin and Turner 1986).

In this book both theoretical inquiries of Strauss and Glaser are incorporated in the study. First, concepts are identified, and their properties and dimensions are discovered in the data. The core task is to dimensionalize the properties of a category (i.e., the dimension "silent to nonverbal to vocal" response signals for the property of pilot-controller interaction). Second, theoretical understanding (sensitivity with some "distance from data") is accomplished with opening up the data which is what the responders themselves are saying (for Glaser) and with questioning and analytic tools (for Strauss). Taking this typology into account, this study starts with the core knowledge concept of aviation safety; category properties and relationships are described through dimensionalization, with theoretical understanding from the responders and questioning tools to codify their interactions. It is what Glaser calls the "chief concern" of the people in the substantive area. Furthermore, in aviation as a service industry and organization, safety is also the chief concern of all the others who are the users of the aviation system and, therefore, is prioritized in research. Variations and categories with properties are constantly revisited during the process of comparing, coding, and analyzing the empirical data. That is why the process of examining raw data (reading through for phenomena), moving to concepts (through sorting) and to preliminary theory (through coding), and finalizing theory is not linear (Martin and Turner 1986).

In this book the core knowledge concept is the central phenomenon for categories to be integrated, i.e., aviation safety via its communication concerns. A category is a classification of concepts grounded under a higher order (like silence and voice). These concepts are discovered by comparison of one against another and appear to fit to a similar phenomenon. Concepts are meaningful labels on discrete occurrences, events, and other instances of phenomena with properties. Properties are attributes or characteristics in these concepts. Lastly, a causal category (like situation awareness) includes the events, incidents, and happenings that lead to the development of the core knowledge concept.

Two major categories that consolidate prior concepts and properties emerge from the data sources, regarding the core concept of human communication in aviation safety. The first is the concept of "silence," as a phenomenon of institutional communicative behavior between aviation actors in the flight deck (with controllers and the airline operator from the outside) and the interpersonal interaction in the cockpit and cabin crew or across cockpits. The second concept is "voice" (vocal channel) as the complementary communication pattern that uses verbal communication (language and phraseology) and administers an interaction with interruptions and truncated messages which needs completion and confirmation.

In a certain overlap in between those two conceptual categories, two other communicative behavioral patterns (as concepts) were discovered in my material and interviews: the non-verbal airplane-to-airplane (and cockpit-to-cockpit communication) and the phenomenon of hesitation in interrupted talk, as vocal but not verbal communication pattern (noted with dash lines in Fig. 4.2, in Sect. 4.3.3), as a vocal channel including truncated messages which to be monitored for the completion of the interaction. Several properties of those concepts unfold when responders comment on scenarios (on Chaps. 5 and 6).

4.3.2 A Causal Category of Knowing: Situation Awareness

Both conceptual categories of silence and voice have a significant impact on situation awareness (SA) of all the actors in this GT approach. Situation awareness is a causal category with a crucial effect on aviation safety. SA (found also as "situational awareness" in literature) was formally introduced in human factors, critical systems design, and cognitive engineering literature by Endsley (1988) while evolved in dynamic systems, weapon/missile systems and safety engineering, healthcare, and humanitarian logistics (Adams et al. 1995; Endsley and Jones 2012; Endsley 1995; Leveson 2012; Dalrymple and Schiflett 1997; Cozzolino 2012). During World War I, Oswald Boelcke, an "ace" German pilot, seemed to identify the SA concept by realizing the importance of "being aware of the enemy before the enemy does so" (Garland et al. 1991; Shaw 1985).

As Woods (1988) noted, the essence of SA is the separation between the understanding of system status by human operators and the actual system status itself. Researchers and military officers in military administration and systems of command and control also argue that "situational awareness" refers to the 1970's sports psychology and has an implicit immediacy (Hone et al. 2006). In Hone's argument, "situation awareness" is a similar (related) construct, but, as linked with Endsley's work, it involves slower processes and removes immediacy. Endsley's model and definitions (1988, 1995; Endsley and Jones 2012) have some discrepancies, but they are still the de facto standard approach for the study of SA. The use of the acronym is more accommodating, and this analysis follows the three types of SA for cooperating actors with the time dimension and the original definition for the state of the individual actor.

From Perception to Cognition SA relates to individual's perception of the "here-and-now," while this state adjusts in time. In Endsley (1988), SA includes perception of the elements of a systems environment (who is where in the flight), comprehension of their meaning (what are they doing during the flight or have done preflight), and projection of their status in the future (what they will do as they navigate and aviate).

In military context, on the other hand, research challenges Endsley's definition by dividing it in parts, in order to add information processing tasks prior to the performed tasks (see Freeman and Cohen 1995; Mavor et al. 1995). In Fig. 4.1 a synthetic combination of different viewpoints (representing approaches from Endsley 1988; Freeman and Cohen 1995; Hone et al. 2005, 2006) and types of SA is pictured:

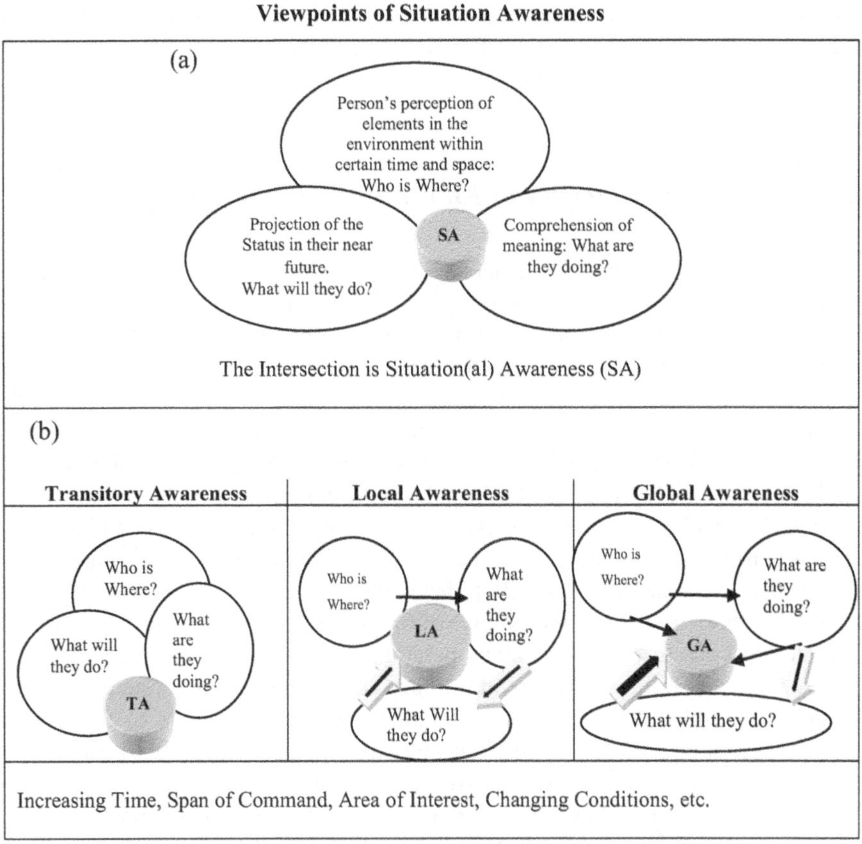

Fig. 4.1 Typology and deconstruction of the concept of awareness (SA, TA, LA, GA) (**a**) A single individual viewpoint- implicit in the original Endsley's definition (**b**) Actors with common interest, the same high-level task, but different data inputs and degrees of three types of awareness

I will guide you through the representations in Fig. 4.1which use pseudo-Venn diagrams with circles on the basis of "perception, comprehension, and projection" which result to "awareness" and determine "cognition." The figure is divided in two parts (a, b) separated with the dashed lined. In the upper part as marked with (a), Endsley's definition is reduced in three questions of operational nature; each question connects (using the short arrows around the circles) with the respective function of the actors involved.

For example, "who is where" includes actors, objects, and features of the environment of the situation. The functions of perception, comprehension, and projection are sequential and parallel, and "awareness" starts at a local level (as current knowledge of task-related events), continues at a transitory level (as mere awareness moving from one situation to the other with turn-taking in communication), and becomes global (as assessment of the full scope picture of flying, controlling, or maintaining an aircraft). Awareness, in part (a) of the figure, is a variable state, an attainment from the answers in the three questions (which in turn describe the phrases from Endsley's definition). In this model, the three questions are considered to be components of awareness, and their answers are necessary "to meet" in the intersection of the three circles. Each component is a stage that requires some data from different forms of information processing. Data input, processing, and reasoning all overlap (as discrete items and as a continuous flow) to lead to SA.

Air traffic control and airfield ground movement of aircrafts, as well as most forms of football, basketball, or disaster management deployment, are activities to apply this model; i.e., the situation is in fluctuation and where all data, evaluations, and assessments are of a transitory nature (Hone 2003; Hone et al. 2005). That is why there is a shift to part (b) in the lower part of Fig. 4.1. The combined answers to the three questions in the circles of part (a) formulate the transitory awareness (TA) in the bottom left-hand side part of the figure.

Two other types of awareness complement the viewpoints of the concept: the local awareness (LA) and the global awareness (GA), in the middle- and right-hand lower side of Fig. 4.1. Those two diagrams of circles contain a separation with an increasing amount (smaller circles of LA and GA). The decisive factors for this separation are the constrained time of limited availability, the span of command (the number of people who report/communicate to a commander rather than the commander's rank), and the geographical area of interest (or the medium, like air navigation or sea cruising). The arrows linking each circle indicate a slower process for memory and information processing, so that "awareness" would embrace locally situated and then global events.

For example, in flight of commercial pilots in a two-person crew, each pilot has a different LA; the pilot-in-command who is the flying pilot will have inputs from the controls (and the actions performed) which may not be available to the copilot. The rate of change of cues in the environment and the key items of information determine measures of awareness. A pilot in a ground attack role experiences a far greater change to the TA than a commercial pilot approaching to land, although the LA situation may be quite similar. Thus, data that are of immediate importance in

TA may evolve as of declining importance to contribute to LA. Any variation in importance is also associated with the changes in operating and flying environment; it is a principle accented in the work of ecological approach to perspective (Gibson 1979).

As prompted below the dashed line in the heading of Fig. 4.1, Endsley's inclusive approach to SA implies a single individual's viewpoint and is deconstructed in three types or stages. When SA is conclusive for more actors involved, then the typology of TA, LA, and GA is a closer-to-reality description from multilateral points of view. In this sense, the air traffic approach controller and the pilot of an aircraft on final approach have a common interest, and they are both concerned with the same high-level task, but they attend different data inputs and acquire different degrees of TA, LA, and GA. Furthermore, when the aircraft is in ground movement (like in Sect. 6.4.3 or 6.4.4), TA, LA, and GA measures change again. In the Helios 522 flight, the two fighter pilots intercepting the silent passenger plane possess TA (after their briefing) and LA but low GA, where as the Chief of HAF (in command post) had a high TA and the Prime Minister had high GA but limited TA and lack of LA. The Prime Minister has the ultimate decision privilege to authorize a shoot down of an "unidentified" aircraft as it was happening with the Helios 522 flight. The pilot flying in the interception communicates his LA to increase the TA of the Chief of HAF, who in turn conveys the messages to the Prime Minister who has to rely on the communication flow of subordinate actors. The decision of "whether it is ever permissible to weigh life against life" and the consequences of a defense of necessity, in this case, are described in several oral accounts of Helios 522 judicial proceeding, discussions with HAAISB members, the fighter pilots who have experienced a Renegade alert, and oral accounts and televised interviews used in this study (see also Tsolakis 2013).

In the typology of awareness in part (b) of Fig. 4.1, the transitory type is closer to the traditional conception in sport psychology, while the local type approximates a closer relationship to the Endsley's model; the local and global types are linked with situations of multiple levels of hierarchy of take-command-from and report-to (like in Mavor et al. 1995).

Assessment of awareness goes with actor's performance for tasks performed, with interference to tasks or interruptability and completion in a certain time frame of availability. The separation of SA in types denotes that an individual could possess awareness of any type (or of all types at once). It is a matter of real topical situational factors (in flight, combat, or other activity) to indicate which type (or types) is relevant. An important warning that Endsley states regarding the assessment of SA is to avoid the tendency of measuring "the participants' confidence in their situation awareness" rather than their "actual situation awareness" (Endsley 1994).

SA, especially in the trifold expansion of the three types, implies clear communicative dimensions as human actors should act upon, respond to, and confirm/correct important informational cues. Apart from that human communication forwards the information flow functioning in LA type and facilitates the move to TA type. The GA type needs also reporting and critical evaluation and access to privileged operational and mission information. Awareness (of all types) is internal to the individual and does not reside on any (technological) information display. The individual actor develops awareness when considering the interpersonal interactions and relationships with relevant communication. Interpersonal here includes the form of the "individual-group-team" or the form of "own vs opposing force/group" in command and control contexts.

Examples of SA Realities The high level of standardization in aviation is expected to streamline SA, but the operating environment still has humans with multiple languages and mindsets working together and interacting in international flights (in diverse conditions and unexpected events). The deadly Tenerife disaster at Los Rodeos Airport on 1977, where Spanish controllers communicate with a US flight and a Dutch flight and crew members, is a clear example of how language and its semantics (the meaning) and pragmatics (the context that contributes to meaning) had an effect to LA and TA types of awareness. The utterance "we're now at takeoff" from the Dutch copilot was not understood (as being in the takeoff process) by the Spanish controller who replied with a vague "OK," after the Dutch captain powered up and said, "Let's go" (since he mistook the meaning of "clearance" at that specific point in time). Then the controller called for "standby and wait for clearance" and increased complexity in the interaction (see Roitsch et al. 1978; Cookson 2009).

The copilot announced that the aircraft was taking off, yet the tower understood his comment as "we're now at takeoff *position*" and not in the "*process* of taking off" (Krock 2006). The eight minutes of interaction escalating to a disaster was time scale with dynamic flux in TA and LA types from one utterance to the other, as the controller had a change to the GA type due to the ground movement of the US flight (also). Then the controller speaks simultaneously with the US flight crew, and the crew members in the Dutch flight find it hard to hear. Neither the copilot nor the flight engineer, in the Dutch flight, questioned their pilot who had a very senior position (as it happens in several responses in Chaps. 5 and 6), and the impact occurred about 13 s later.

Derangement of all types of awareness for the different actors, in this disaster, was combined with additional factors all contributing in the process for "memory and information processing" of these knowledgeable actors (as indicated with the pointers linking each circle in Fig. 4.1, part (b)). The fog in the weather, the smaller and less familiar aerodrome, the news of bombing Gran Canaria International Airport (which was their original destination), and the intended expedition and rush to cover the schedule delays from previous days were all in the background of that disaster (and the TA and LA of the US and Dutch flight crews and the GA of the controller).

Rescue, disaster relief, and military operations are now increasingly multinational and multi-force and therefore with increased potential for misunderstanding. For example, under the command of NATO or the UN, it is quite possible to have an English-speaking commander whose orders are mentally translated into (let's say) Dutch or Greek by his chief of staff, further mentally processed and translated back into English, to be passed to a squadron commander, namely, French or

Turkish, who repeats the double translation process to brief his Italian subordinate and his Norwegian deputy commander, a woman who served as a liaison officer with the local community (as it was happening in Cyprus in the United Nations Peacekeeping Force).

The relay of commands and the confirmation of information flow is a demanding communication accomplishment which has an effect to awareness of LA, TA, and GA type, in a different way for these actors that need to know and decide "what to do." A similar situation of interpretive messages fragmented information flow, and gradual development of awareness happens in NATO reconnaissance flights with multination crews of the NATO AWACS (like the pilot who responds in Sect. 6.4.1). As Hone (2008) points out, such coalition processes in command and control environments generate a major communication challenge (of language and culture) that clearly affects SA (and consequently effectiveness and safety) especially in the immediacy of a flight situation.

In the emerging causal category of SA in this study for this book, four key elements cooperate: humans, informational cues (as stimuli processed), behavioral cues (paralanguage and role expectation attributes), and appropriateness (as regularity) of the responses. This regularity implies the comparison of the response with an anticipated response (a choice from many possible expected responses). These expected responses formulate an instrument for a performance measure of SA, from both pilot's and controller's sides, as it is apparent in the responses analyzed in Chaps. 5 and 6. For example, for a military pilot like those in HAF, SA (TA type) means to understand the threats and intentions of the "non-friendly" aircraft or enemy forces, as well as the status of his own aircraft (LA type). In order to project the situation on his near future, it is inevitable to communicate from the start (communication builds a foundation for a GA type also). For an air traffic controller, SA denotes (at least partly) to estimate about current aircraft positions and flight plans and the required separation between these positions and entities (GA type), in order to predict future states of conflict (LA type). Of course, the controller cannot be a simple observer, but communication is crucial in all stages and has a considerable effect to the SA of the cockpit crew (TA type for the controller). Thus, situation awareness, as an emerging communicative parameter, includes acquisition of information from the immediate (and the co-present controller) surroundings, integration with relevant knowledge of the situation, exploration of a mental picture for the situation, and anticipation of reactions for the immediate future. Hence, situation awareness has a clear effect on aviation safety and should become an added component in communication models when the sender and receiver should conclude on "how to know what to do."

4.3.3 The GT-Emergent Knowledge Model

The theoretical constructs of concepts and categories, using comparisons and sorting from empirical data, conclude in an abstract figure which represents the emerging knowledge theoretical model, in the scope of the current study. Figure 4.2 below shows the theoretical array of concepts (in the above typology) emerging from this study using a GT approach:

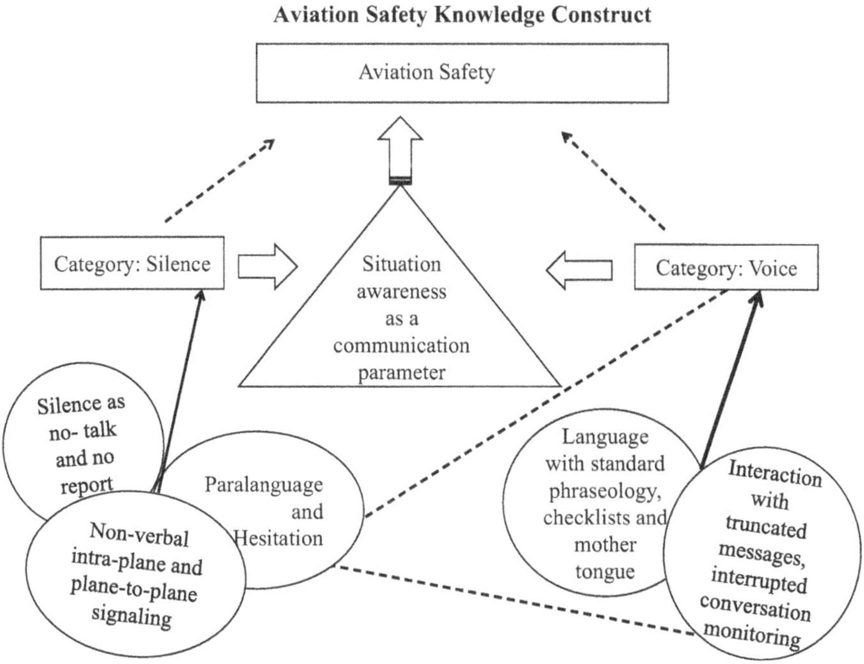

Fig. 4.2 The core knowledge concept under study with interrelated categories

Two main parts position themselves in the two branches of the figure as categories related directly with SA (described in Sect. 4.3.2). Let us go over the structure and reasoning of this model.

Silence Category On the left bottom part of the figure, there are three interconnected circles containing concepts with properties identified in the data from flights. The first one is silence in the cockpit in the form of no-talk: not confirming/reading in checklist marking between pilot and copilot or using a silent checklist to mark, not replying to calls from air traffic control for extended periods, no talking between cockpit and the cabin, standing-by silent in an expected interval till the next phase of flight, or in the form of not reporting flight incidents in the flight log or in the briefing/debriefing.

The second circle contains the concept of nonverbal communication of human actors across different airplanes and cockpits. It could take the form of waiving hands across cockpits; waiving wings during flight towards another aircraft; approaching the other in proxemic distance; using territorial markers in a cockpit to signal operating status; and activating silent and audible but visually mediated messages in banners and symbols from the cockpit to the passengers or the form of radar locking in a fighter jet dogfight. All these attributes are different ways to signal a message in nonverbal channels.

The third circle contains an important vocal but still nonverbal concept describing the component of paralanguage in the form of nonverbal cues of the voice or mainly hesitation talk (that is rather truncated or not). Hesitation may convey emotion or nuance meaning with forms like prosody, pitch, volume, and intonation. Hesitation is commonly accompanied by silence, as a result of uncertainty for the receiver or a reluctance to correct from the sender. Hesitation is closely related with meta-communication, as it communicates about other already transmitted messages.

All circles in the (pseudo-Venn) diagram have a common part of interconnection, as usually one complements the other in an interaction as communication components. Per a GT approach, we can combine (with the arrow pointer) all the concepts under the category of silence in the parallelogram above labeled "silence." Silence in turn has a causal effect on SA, which is crucial to aviation safety (both relations are represented with a thick arrow pointer). Rarely, but with fatal consequences, silence may have a direct effect to aviation safety without passing through the SA of different actors (like in the case of Helios flight crash). In such an instance, silence emerges as a causal category by itself, usually as a safety threat for what is known for the flight, represented with the dash arrow directing to aviation safety parallelogram.

Voice Category Similarly, on the other side of Fig. 4.2, we have two other communication components that are conceptualized for this GT approach under the label "voice."

The first left circle from the right side of the figure represents a major verbal communication channel with several attributes of language. Standard phraseology is expected in message interaction, in the form of aviation English (in international flights) of the local language (in domestic flights); most checklists are specially designed verbal messages that are pronounced in a dialogic format (copilot asks, pilot acts on controls, copilot marks the checklist) to ensure collaboration and avoid mechanical parsing of the checklist. Language may also take the form of mother tongue, even as an exception from standard phraseology. The most common (and justifiable) instance according to the responders are emergency, constrained time or unusual situations, or laconic but metaphorical interactions during fighter jet pairs cooperating in a dogfight against intruding aircrafts.

The second circle in the right side of Fig. 4.2 contains verbal communication, rather fragmented, really applicable in real-flight situations where the sequence of flight operating manuals does not represent the interactional realities. Messages are truncated messages to save time, change turns, or continue a rapid pace; messages may be interrupted by other incoming inputs or change of communication channels (e.g., stopping to converse in one frequency, asking for a clearance while a control checking is conducted, aborting a readback/hearback because of an occurring event or a request from another flight). In most of those times, a reasonable effort for a standby rather interleaving mode is needed (from the cockpit for their own tasks) and from the controllers (to catch up the next flight's request). A monitoring process to "recover" the conversation or complete the truncated messages with new messages is needed. A process that is closer to meta-communication as original communication was interrupted, and the new messages convey meaning about the old ones.

Again, the two circles in the right side of the figure are represented with an intersection (in a pseudo-Venn diagram) to show that they are not mutually exclusive but co-exist in the data and in the context of verbal interactions and they are combined under the category of voice. Voice, as a category, in turn has a causal effect on SA, which leads to aviation safety, as pointed out with the thick arrow to the left. In less common instances, voice may direct its impact to threaten aviation safety or enhance it by mobilizing a controller who shouts when looking outside of his window for the actual position of aircrafts in the runway. In these occurrences the effect of voice goes directly to safety, making it a causal category itself (like in the case of the controller in Thira Airport of Sect. 6.3). That is, the dash line pointer from voice to aviation safety parallelogram of Fig. 4.2.

Lastly, as communication theory suggests, and respondents point out (in Sect. 7.3), paralanguage and hesitation have a vocal aspect and related with truncated messages (Grice 1989). This type of relation that connects the circles of different conceptual categories is represented with the dash lines from the respective circle (hesitation on the left to truncated messages circle and the voice category on the right). In this book, a GT approach through immersion in sources of data, comparing and coding (in interviews and transcripts), leads to the identification of five different phenomena of communication, which function with interconnectedness (showed with intersections). The first three are nonverbal communication signals, silence, and hesitation as paralinguistic instance. The next two are the use of language and truncated messages recovery which emphasizes verbal communication. Both categories have a causal effect to SA which, in turn, with its pointing triangle in Fig. 4.2 ultimately signifies the direct effect to aviation safety.

Aviation safety is the desired result – i.e., to know how to accomplish it – when moving people by airplane to a certain destination and includes investigation, categorization of flight failures, and the submission of prevention mechanisms to avoid for such failures. Aviation safety takes the form of a research question with the prompting of human communication as a key contributing factor with impact in aviation safety. That is the starting point of a GT approach with a certain knowledge preparation in literature of human factors, systems theory, and communication in order to locate and select appropriate data sources, informants, archives, and raw data from in-depth interviews and filter them through iterations with real-life practitioners.

4.4 Data Processing and Harvesting Knowledge: GT Sorting and Coding

Before reaching the point of the actual scenarios exemplified in Sects. 3.1 and 3.2, a process of sorting and coding is necessary to identify relevant questioning topics and discover emerging concepts to build constructs. In this next section, an example of a coding process using data from an accident investigation report presents one of the applicable coding/comparing techniques, to explain theory construction "about

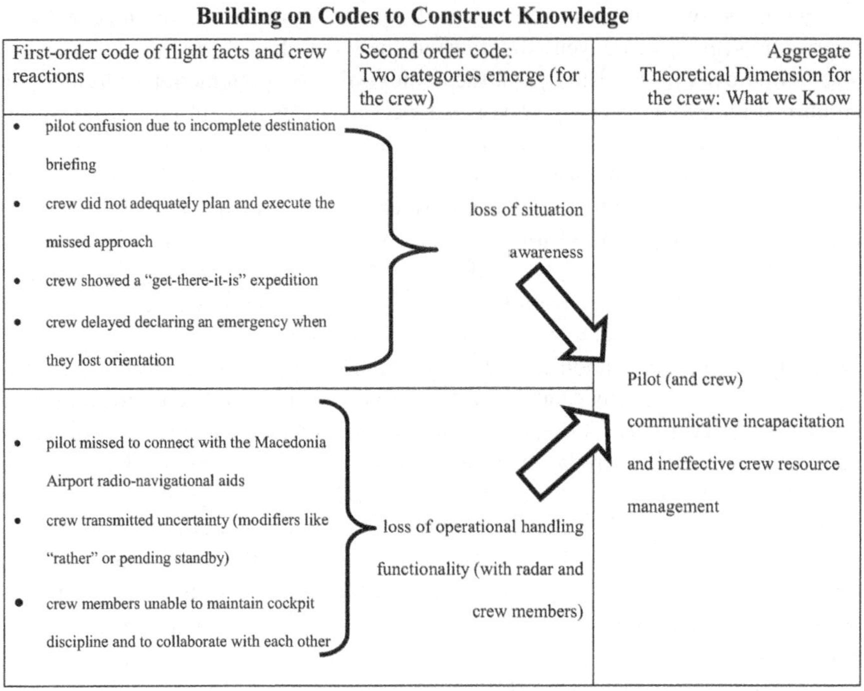

Fig. 4.3 Example of constructing the theoretical dimensions of "Pilot Incapacitation and of Ineffective Crew Resource management"

what we know" better. As explained in Sects. 4.3.1 and 4.3.2, in GT methodological approach, the core task is dimensionalizing category's properties (i.e., the dimension "silent to nonverbal to vocal" response signals for the property of pilot-controller interaction), and opening up the data is what the responders themselves are saying while extracting concepts with analytic tools (for Strauss). As pictured in Fig. 4.3, the parallelogram with coding comes after (and revisited) the phase of initial sorting and comparing in bottom right-hand parallelogram. Variations and categories with properties are constantly revisited during the process of comparing, coding, and analyzing the empirical data. In this section, a structured coding process is explained with an example, as related to the core concept of aviation safety explained in from Sect. 4.3.1. The start is from an accident report, to showcase a source different than the interview transcript in the GT approach applied (with the resultant concepts).

During the visit in the headquarters of the Hellenic Air Accident Investigation and Safety Board in old Hellinikon Airport in Athens, their emphasis was in improving the multidimensional communication and augmenting standardization in language use. HAAISB's personnel have agreed to serve as partners to evaluate other participant responses as per the concepts I am categorizing via the initial questionnaire. They also provided several reports from their accident investigations with transcripts and their institutional comments. The study of these transcripts allowed the identification of several codes and concepts to emerge, as well as side notes to evaluate the communicative significance of these factors and a route to connect with other parameters of the flight.

Using an Accident Report The accident report of the Ukrainian (AeroSvit Airlines) Yakovlev YAK-42 crash of December 17, 1997, in a mountain outside Thessaloniki in northern Greece is the primary resource for this example (HRMTC 1998[1]). For this phase, where information is based more on testimonies of the flight and judicial proceedings, an initial coding is oriented to operational attributes of the flight. It also incorporates several instances as observed before or after the flight. A summary from the accident investigation report follows:

> Flight 241 started at Kiev with a Boeing 737; engine problems required an aircraft change at Odessa, with a Yakovlev-42. After a missed landing approach (ILS system) to runway 16 at Thessaloniki "Macedonia" Airport, the crew was instructed to proceed to the North Hold for a second attempt. The airplane did not follow the published missed (ILS) approach procedure and headed West-Southwest, flying into the side of Mount Pente Pigadia at 3300 feet, declaring an emergency with a considerable delay. After the last call on the direction from the Control Tower the pilot replied, "I will report" but no reply is recorded after the truncated message. The wreckage (of the crash) was found at 10:30, December 20, due to the thick snow and harsh weather conditions with seventy-four fatalities.

In the court trial, two air traffic controllers were accused of many counts of manslaughter and of violating the designated transportation regulations for the use of radio facilities in that airport and for the delay in responsiveness to the crew questions. The final verdict of the appeals court on 2002, for the two air traffic controllers, was a sentence of 4 years and 4 months in prison each.

By reading the transcript and the HAAISB's accident investigation report, a first-order coding and relation conceptualization is accomplished in a GT approach in a diagrammatic technique to represent the knowledge acquired, as proposed by Grover et al. (2014). Figure 4.3 that follows exemplifies a process of building on codes to construct theory, the knowledge from the Yakovlev crash:

I will guide the reader through the three-stage process in this example to follow argumentation for "what to take out" of this case. The first column on the left-hand side of the table records the flight facts as recorded in the flight recording devices (CVR and FDR), the transcripts of crew-controller interaction, and the analysis of the accident investigation team. Charmaz's typology (2006, 2009) of

[1]Air accident investigation in Greece was carried out by the HCAA itself after World War II and then by a committee reporting to HCAA and an independent council. The Hellenic Aircraft Accidents Inquiry Council was operating till the establishment of a functionally independent Air Accident Investigation and Aviation Safety Board (AAIASB) in February of 2002 (the legal framework of HAAISB is described at http://www.aaiasb.gr/en/aboutus/istoriko1en.html and EU Regulation 996/2010 at http://www.aaiasb.gr/images/stories/documents/R%20996_10%20EN%20Accident%20Investigation.pdf). USNTSB was consolidated on 1967 and reestablished on 1974 as a completely separate entity (as described at https://www.ntsb.gov/about/history.html). ICAO (from 1951 to 2001) standardized the procedures for the investigation of aviation accidents and published Annex 13 of the Convention on Civil Aviation on "Aircraft Accident & Incident Investigation" (ICAO 2001).

coding questions recommends an inquiry for the goal of study, the pronouncement from the data, the point of view of the theoretical category, and the consequences of action from the participants. By evaluating these inquiries, the coding process moves ahead.

So, by unfolding backwards the safety risks discovered in this flight, crew reactions are categorized in the following concepts: pilot confusion due to incomplete destination briefing, insufficient planning as crew did not adequately plan to execute the missed approach in certain weather conditions; with their cumulative workload of changing aircrafts and transit destinations, crew showed "hurry-up" expedition to land as soon as possible in order to be able to return back; and crew delayed to declare an emergency after they missed the first approach although they expressed an orientation problem (the controllers did not seem to grant the required priority due to the delayed declaration of "emergency"). These reactions from the crew are combined (with the bracket in the margin of the second column) to the pattern of losing the SA, the sense of "here-and-know," and the next anticipated action.

In the second cell of the first column in Fig. 4.3, crew reactions that relate to operational handling are grouped. The pilot missed to connect with the Thessaloniki "Macedonia" Airport radio-navigational aids that are complemented by a military radar for airport approach, a possibility that was not clear in their briefing and was not explicitly suggested by the controllers. The transcript shows several hesitation or uncertainty moments with messages of modifiers like "rather" or pending "standbys" instead of definite qualifiers in sentences. Cockpit voice recorder monitored some tense talking and disagreement between crew members who were unable to maintain cockpit discipline and to collaborate with each other. The grouping of those reactions with an operational nature leads to the category of loss of operational handling functionality, as if the pilot and crew could not operate (navigate) inside the cockpit. The second bracket down to the margin of the second column shows the relations of the first-order coding phrases with the category which emerges.

The second column of the Fig. 4.3 connects (with the two thick arrows) the selection of coding phrases (of the first column) with the second-order codes which correspond to categories of incapacitation: no situation awareness and no handling capacity. In turn, those two categories aggregate in the third column (with the two pointing arrows) as causing communicative incapacitation and having a result of ineffective crew resource management. The interaction with the air traffic controllers did not accomplish a cohesive cooperative mode, as the crew seems to have a slight intercultural gap and did act efficiently inside the cockpit; i.e., they had trouble communicating as their handling was inconsistent to the circumstances, and other crew members did not have a positive contribution.

The coding scheme explained in Fig. 4.4 allows us to locate another rather latent conceptual dimension in the representative nature of situation awareness. As Flach (1995) points out, a form of circular reasoning might occur when SA is overestimated as an objective cause of flight accidents: "How do we know that SA was lost? It is because the human actors responded inappropriately? And why did the human respond inappropriately? It is because SA was lost." In order to avoid tautological conclusions, SA in this study is connected with human communication consequences/effects to aviation safety. The process of SA refers to

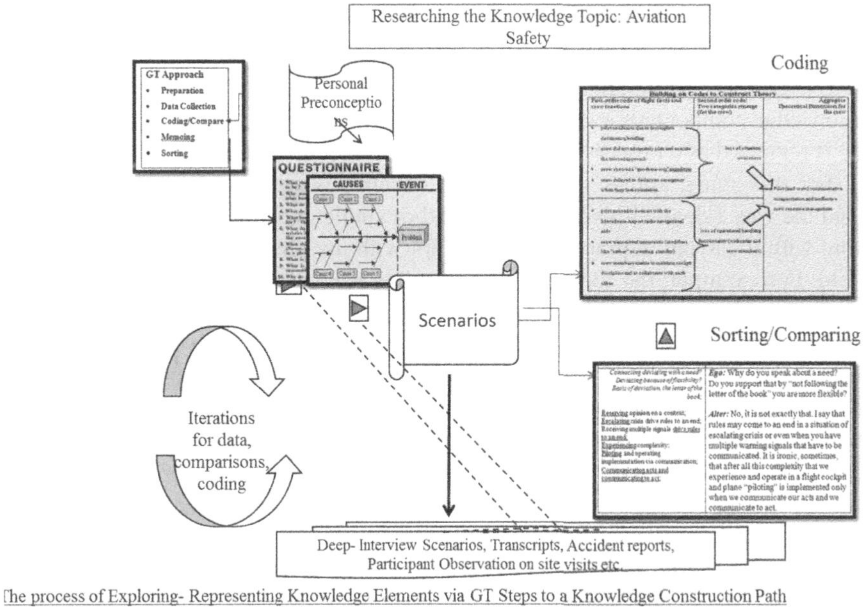

Fig. 4.4 Reviewing the process of exploring knowledge using a GT approach in this book

how SA develops and sustains during flight. At the same time, SA is the resultant concept, like a product of the process. The coding process was not just a simple application but allowed me to uncover conceptual variations, as noted above, that were useful in configuring relations among categories in this exploratory study (as explained in Fig. 4.4).

This is an example of reading data with coding-analyzing, as processed from my reading of an accident investigation report. It is an additional source, before the interview questionnaire to increase applicability of the approach (Glaser and Strauss 1967). The inquiry developed the following three questions for further sorting/investigating in the volume of data:

- Is it a case of miscommunication due to the eastern/western cultural split and incompatibility or just an issue of linguistic standardization?
- Is it a human navigation error or a "non-listening the controller" event (when the delay of hearback and instruction-giving from the controller might be fatal)?
- Is there a (electromagnetic frequency) gap in the airport maps with the radio navigation devices?

For another time an accident is attributed to a combination of (otherwise) minor events which accumulate with a communication breakdown situation. Communication breakdown occurred across several dimensions, and the crew members were not effective in collaborative repair.

4.5 Representing Knowledge in the Flow of a GT Approach

A GT approach is iterative through theory, data, concepts, and constructs of the knowledge components regarding aviation safety. In this section this kind of dialectic is reviewed as a process of acquiring the knowledge of lessons learned.

For this study, author's personal preconceptions are used from the start, and lead the way I deal with scenarios instead of single questions, to make them relevant with the aviation practitioners and not with my preconceived ideas about the field. For example, the idea that fighter pilots have a mindset of urgency and of "living the moment" qualifies a questioning process that is relevant with their work and highlights their main interest of aviation safety tight with mission efficacy. Also, in another idea, for communication directness in airport activities where the content of the message must be expressed in low-context language, the questioning process requires self-sustained scenarios which point out to the direct topic. A great deal of relevance was accomplished after a series of visits for evaluation and active observation in aviation installations, as well as from the accumulated participant observation and work in such environments. A process of constant comparison starts with the first level of reading and identifying phenomena of interest in the analysis of data, and then a decision on the theoretical sampling is taken; it is about what type of data are to be collected next to support the theory construction (Glaser and Strauss 1967).

A schematic representation of the GT approach as applied in this study in the search of the knowledge theme of aviation safety is depicted in the following Fig. 4.4:

In this book, the GT approach involves identifying the theoretical and empirical core concept of aviation safety (Sect. 4.3) and selecting the data resources (multiple times in Sect. 4.2). The process is iterative (with comparisons and coding) at all stages as represented by the thick circular arrows in the bottom left of the figure. This iteration is a constant follow-up at any stage since it facilitates the cross-concept comparisons. Also, after the preparation and initial data collection described in Sect. 4.2, coding, comparisons, memo note-taking, and sorting of information for concepts to emerge were in the list of a GT approach for this study, as shown in the top-left short list (labeled GT approach) in Fig.4.4. The item "coding and comparing," which is pointed out with the arrow in the list, is the main concern to allow "aviation safety" and its causal categories of "situation awareness, via silence and voice," to disclose and build up

Preconceptions Orienting the Steps Next to the right of the GT listing in Fig. 4.4, a banner of "personal preconceptions" is drawn to initiate the questioning inquiries. From my previous background with aviation, I have internalized three categorical experiences which had a radical impact in my theoretical mindset and conceptual thinking regarding "what I knew" about aviation and "how knowledge emerges" for aviation. Along these lines, three preconceptions oriented the next steps.

The first idea has to do with the denseness of aerodrome time which is full of events and expected arrivals, departures, and occurrences for all professionals in its functions. The second idea is that fighter pilots appear to belong in a distinct category of service of a different kind of competency, with standards consistent with Rdeinhart (1970). That means that their understanding of what is important has a scale from "a mission accomplished till a moment of silence staring the horizon"; i.e., thinking about the basics in order to realize the essential in the moment, in life, and in mission. In these terms, civil aviation pilots and controllers have sometimes difficulties to connect (and communicate) with such a mindset. Several studies are consistent with the manifestation of such difficulties (Connel 1995; Dismukes et al. 2007; Retzlaff and Gibertini 1987; Ashman and Telfer 1983). The third idea is that communication directness in aerodrome activities goes hand-in-hand with professional responsibility and mutual trust. Mistakes are made, and unexpected events happen, but should be recognized and clearly corrected/resolved and used for coaching and self-improvement (as in Dekker 2010).

Three-Stage Questioning Route The next step going down in Fig. 4.4 with three document layers corresponds to the three stages of a questioning route that was part of sampling for theory to find applicable formats and representative concepts. As pointed in Sect. 4.2, a questionnaire was the first attempt which was dissolved due to the lack of discussion focus; then I introduce the Ishikawa (fishbone-shaped) diagram with contributing "causes and a key effect (in the fish head)" with a different core topic regarding the conformity and deviance to rules. The operation-related fishbone format disqualified its use from commercial pilots and controllers, whereas fighter pilots where keen on it. The Ishikawa diagrams (also known as fishbone diagrams or cause-and-effect diagrams) are visual tools that show the path for contributing factors to a certain event (area of interest or problem). Ishikawa diagrams were first introduced in quality improvement of industrial processes (Ishikawa 1968). The diagrams display the factors that affect specific quality characteristics or problems (Tague 2004; AOTS 1985). Typically, a fishbone diagram matches a process of brainstorming (Nelson 1998), used in this study in the second stage of data collection, regarding the topic of knowledge and rules. A fishbone diagram is one of the first steps in investigating a field of interest and includes people involved, processes and requirements, assessments, categories/questions, and the environment. In aviation industry this tool was used to investigate the quality of air travel service, among other topics (Newby et al. 1996; Smith and Ragan 1993).

A GT approach recommends the appropriate tool for the greater amount of data and the search of common conceptual indicators in interviews and observations. Hence, the core knowledge concept of this study shifted from rule deviance and tacit knowledge to aviation safety via communication. The qualified tool, in the third stage, was the collection of scenarios/exceptional cases of real-flight situations which encourages the responders to offer an opinion closer to a reenactment of their own functions, practices, and experiences and not as a predictable (by the instruction manual) response.

Each questioning tool was sorting out topics and was orienting the study closer to human communication and its effect to aviation safety, despite other operations, cultural, technical, and commercial parameters that came out from the first two stages of the interview process. That is why the arrow buttons in Fig. 4.4 (from questionnaire to Ishikawa diagram to scenarios) are pointers in a road map. The scenario collection became more representative to concepts as more expert responders were recruited to achieve consistency in the content and the relationships with the concept categories of "silence" and voice." The responses to those scenarios disclosed the observed complementary character of the five communication components of no-talk, nonverbal communication paralanguage of hesitation, language, and truncated interaction. These concepts complement each other in aviation interactions and generate overlapping attributes in the main categories.

In the lower bottom foundation of the Fig. 4.4, I have identified the main raw data archives used from the main sources of Sect. 4.2 (in a textual representation of file forms): deep- interview scenarios, transcripts, accident reports, and participant observation on site visits. The dashed connecting lines to the first two stages of questioning tools represent the process of consulting data and sources in each stage before constructing a new tool and selecting the appropriate topical scenarios. Appendix A, of the current book, includes the questioning tools in the three stages and the full collection of scenarios that allow this study to expand further in the next stages. Two examples of scenarios are introduced in Sects. 3.2.1 and 3.2.2. Also, an example of a coding scheme using a different type of data (from accident investigation reports) is presented in Sect. 4.4.

Sorting, Comparing, and Coding On the right-hand side of Fig. 4.4, two snapshots of the processing of a GT approach are represented: one technique of coding from a first order of codes (identifying properties in the data) to the second order of coding (structuring concepts) to the third-order coding of an aggregate dimension which emerges as an evolutionary stage of the structured concepts (from left to right, with another manifestation of the SA concept, in that flight). The label "coding" in the parallelogram used in the process is applied in an accident report with secondary (and not raw) data from accident investigation, combining a transcript and investigator's comments. The full example of this coding is described in Sect. 4.4 for the Yakovlev-42 crash. The second square-like frame below the coding parallelogram is labeled "sorting/comparing" as it corresponds to the close reading/open coding process of a first-stage questionnaire interview. "Ego" and "alter" interact in the right column, and the actions that are communicated generate coding gerunds in the left column. The interpretation of the questions is italicized (on the left), and labels are selected to preserve language and action in the data (from the raw replies of the right column to codes like "reserving opinion, experiencing complexity, or piloting and operating"). The sorting/comparing function informs the coding function in the upper right-hand side parallelogram of Fig. 4.4.

In this direction, functions of coding, sorting, and comparing are used for constructing knowledge. GT's central function is the emergence of theories

(Hildebrand 2004). GT immersion in the data searches for the following knowledge-construction elements:

(i) The relevant conditions in which participants act in the various flight-related situations (e.g., routine flight, emergency flight, oxygen starvation, external or psychological but communicated signals)
(ii) How the actors respond to changing conditions and unexpected events (e.g., in the case of Helios 522, silence in communication due to hypoxia or waiving wings from the fighter jets to convey a message of interception)
(iii) What are the consequences of their actions or lack of action (e.g., after understanding the autopilot control loss in a rapid descend in the Falcon VIP flight? The pilot or copilot did not activate the seatbelt sign; the purser has been trying to prompt the VIP to sit and buckle; and the pilot had the priority to save the aircraft while the VIP on board crashed his body and died from the oscillations; when he had the option to sit, he did not pay attention with fatal consequences.)
(iv) What is the essence of expertise in critical environments (i.e., the expert complies to the "letter of the rule" and to the "spirit of the rule" or deviates when he/she identifies an exception or exceptional case)?

In Glaser's suggestions (1992) theory construction and knowledge harvesting start from the inquiry on three generic questions in the third stage of questions in Fig. 4.4:

(i) What is the main concern or problem of the actors involved in the flights (how to communicate their situation to and from the cockpit, as described in Sect. 4.3.1 of this study)?
(ii) What accounts for most of the variation (different options, choices, opinions, and decisions, like silence, nonverbal communication, hesitation, and the use of voice with standardized phraseology, mother tongue, and the monitoring of truncated messages in Fig. 4.2) while processing the problematic situation?
(iii) What category or what property of what category does this incident (the observed variation) indicate (silence and voice with the intersection of hesitation and interrupted messaging in this case)?

These questions are floating during the comparisons, coding, and analysis in the iterations of empirical data and correspond to knowledge-oriented inquires. This process of inquiry aims to catch the interplay and compare interactions and behaviors with other relevant realities as expressed by the responders (when we speak about interview data) or described in accident report (in accident investigation documents) or monitored from flight recording devices (in the so-called "black-box" transcripts).

Emergence, Theoretical Sensitivity, and the Coding Paradigm Three other key concepts originate from Glaser's and Strauss's (1967) common ground: emergence, theoretical sensitivity, and the coding paradigm. These concepts were useful during the iterative investigation process and the last coding phase and represent knowledge elements acquired.

A general comparative analysis of raw data would allow the "emergence" of conceptual categories from the data of social research, without forcing the data to fit theoretical literature of the study area. There is a downside in the "emergence" of categories from data using the open-coded strategy of Glaser and Strauss (1967); i.e., referred also as substantive, coding means reading the data text line by line for ad hoc coding may result to mass data in abstract insignificant categories with limited new information (when pairing for comparisons in the "sorting/comparing" parallelogram to the coding parallelogram).

That is the point where "theoretical sensitivity" emerges, as the ability to reflect upon empirical data through the lens of theoretical terms and the perspective to facilitate the identification of more relevant data, step-by-step. In these grounds, I decided the move from the fishbone main inquiry (regarding rules and tacit knowledge) to the inquiry of aviation safety with scenarios, a shift which was more inclusive to commercial and military pilots and controllers as well and allowed me to motivate several higher-ranked professionals to respond.

In this book, the study is using the "coding paradigm," in the "coding" parallelogram, to cultivate concepts that seem to fit the data (Strauss 1987; Strauss and Corbin 1990). The coding paradigm has four items: conditions, interaction among actors, strategies and tactics, and consequences, all applicable to represent knowledge acquired. Coding operates by breaking down of data and developing concepts organized in new ways and builds a skeleton or an "axis" (that is axial coding) to develop grounded concepts. Coding means analyzing but is also a process of putting tags, names, and labels in pieces of data (Robson 2002). It is a useful tool to unpack human communicative behavior, as manifested in interaction, with the potential of pointing out strengths and weakness in the communication process. The goal is to construct a knowledge theory to describe the data and the phenomena that are manifested through the data Charmaz (2006). That is why from both "sorting/comparing and coding," parallelograms dash lines lead back and forth the files of deep-interview scenarios, transcripts, accident reports, participant observation on site visits, etc., in the bottom part of Fig. 4.4.

The exploratory study is the essence of a GT approach. After collecting an initial small amount of data, then theory construction begins for the specific domain. Data may be an insufficient data, and the theory may be incomplete, but the researcher tries to determine its weakest point. A useful technique in this phase is the so-called abduction (Peirce 1958; Reichertz 2007), a process of forming explanatory hypotheses to logically introduce a new idea. Abduction brings out a surprise that opens new insights and the kind of "discovery" that is at the heart of the grounded theory method (Reichertz 2007; Muller 2014). The weakest point of the incomplete theory is the start of a (what-if) condition (leading to a likely explanation) for new data selection. The goal is that by learning how it fails at its weakest point, a large amount of information emerge for making a stronger theory based on that failure (Awbrey and Awbrey 1995). With this process the main topic shifted in the second-stage fishbone questionnaire to the scenarios with the focus on aviation safety and not on deviation from the rules.

4.6 Discourse Constructs Knowledge About Aviation with Close Reading

The iterative character of a GT approach is quite compatible with the textual analysis for notes, narratives, and interviews from specific field visits. Since the collection of materials has the form of text (directly or through my memos), the processing analysis of standard textual information takes the form of a discourse analysis in Chaps. 5 and 6 (Paltridge 2006). In discourse analysis the phenomenon of study is identified (as the relation of human communication to aviation safety), different types of texts as data source are located (from accident reports to interview transcripts here), and tags to label significant features in the texts, as codes, are assigned.

Formal accident reports and operating and regulatory manuals generate the textual framework. The direct contact with pilots and controllers in informal after-hours discussions in their environment is oral information documented in memo notes. Transcripts from flight recordings and relevant judicial proceedings allow comparisons of concepts and assertions. Memo notes from participant observation of actual work-doing on site inform the interpretation of observed interactions. The observance of institutional briefing sessions, the data coding and evaluation of concepts with the participants themselves, and the sorting of concepts and their relations with theoretical communication properties are all documented in text of different style and narrative.

In the analysis of oral and written discourse in aviation, it is important to understand the mechanics of language (and the language of special purposes used in aviation) but how human actors practice "what to say to whom" and "how to say it appropriately in a particular situation." That is the notion of communicative competence introduced by Hymes (1972); i.e., knowing what is correct and what is not (according to aviation phraseology and institutional contact standards) but also when and where to use language appropriately (like the motivating, collaborating language devices, the suspending order to copilots, the laconic calls to control tower, the readback of important instructions, the nonverbalisms that communicate emotion, activate-mediated banner messages to the cabin or message updates to the passenger seats with no announcement) and with whom (controller, copilot, cabin crew members, passengers as a collective).

For example, communicative competence in aviation requires knowledge of the rules of speaking/addressing the others but also knowing the way the utility and expected response to speech acts (Searl 1979). For example, a "*May Day*" call from the pilot should mobilize the controller for search and rescue procedures. An erroneous readback and the resulting clearance violation require immediate correction and apology from the pilot to the controller (even in person, if possible and needed like in Sect. 6.4.2).

Discourse analysis originates in studies of the sociology of scientific knowledge in the context of social psychology (on the one hand) and in the writings of Foucault (on the other hand, 1972/1995). Hence, it is appropriate to analyze interaction of highly skillful knowledge practitioners in aviation, as it happens in this book, moreover, because it involves gathering of data from a variety of empirical sources, formal and informal, as it happens with a GT approach. The method suggests that interactive dialogue generates commonly held beliefs in a sense-making process (Wooffitt 2005; Gilbert and Mulkay 1984; Potter and Wetherell 1987). In aviation, crew members and controllers need to establish their topical awareness, as it makes sense at the specific conditions they are in. This kind of discourse analysis, in social psychology, investigates accounts of utterances which serve some function of a context (Maddux and Johnson 2009).

The coded concepts used as emerged in the study from a GT approach are cross-evaluated with expert members of the participants who evaluate in iterations with me (in this case mostly accident investigators and top-rank Air Force officers). Then the patterns of language across texts relate to each other and to the social and cultural context in which language is used. These stages in the analysis of the discourse are in line with the coding-comparing iterations of Sect. 4.5. A GT approach suggests continuation till the emergence of the two categories (silence and voice), as non-antagonistic but complementary tools in communication and the construction of the situation awareness concept of human interaction in the communication that contributes to aviation safety (in the top part of Fig. 4.4). The nodes in Fig. 4.4 are all communication constructs which arise from the interaction. A close reading process is used to deal with the text with care and a magnifying glass. Parsing the text for the first impressions, the vocabulary and diction, the discerning patterns, the points of view, and symbolisms directed the grouping of instances to aggregate concepts. The expectation is to notice the details and supporting ideas of the textual responses, as they recount the responder's experience or opinion or even criticism from his/her point of view of the scenario (as controller and if one were in the position of the pilot or as pilot and if one were in the position of the controller). Close reading allows us to reflect on the meanings and the order of individual words and sentences and the development of ideas which lead to a thorough understanding of the responder's texts.

Consider, for example, the case of the Falcon VIP flight (of Sect. 3.1.1) in which the cabin crew supervisor felt as if she was attached to an "afraid horse" and the pilot took one step back in replying to the questions to explain the wider picture of the vital risk he was coping with in the rapid descend. In another instance, the experienced controller accepts a bad moment to occur without blaming the pilot, thinking of the collaborative scheme in their work. The fighter pilot in the Renegade alert for Helios had a breathtaking moment of silence (when he witnessed the crash, while it was happening).

The discourse analysis perspective, in this study with a GT approach, also includes the participant actors who perform and receive discourse with cognitive and normative communication devices. Analysis of discursive data proceeds with the detection of "oral evidence of mental constructs," discovering power relationships with ideological stances, rhetorical devices, and cultural expectations (Turner and Warren 2010). That is why discourses are realizations of social phenomena (both in sub-communities of aviation professionals and military officers and for society at large). According to Hall (1997): "discourses are ways of referring to or constructing knowledge about a particular topic of practice." That topic, aviation safety in this book, provides ways of talking about it, knowledge forms, and associations with it.

Language and other social practices are responsible for the production, actualization, performance, and transformation for such devices at different social, historical, and geographical places (in this case the study draws from the aviation sector of Greece, as a leading example of the European and Mediterranean region and NATO alliance, with a dense network of flights and airports, during the era of economic crisis and with the open tensions regarding the Aegean Sea airspace and the exclusive economic zones of the islands).

Discourse is not an entity per se but a set of utterances, a theoretical device for sorting and analyzing data, and a necessary hypothetical assumption to start research in certain time and social and geographical space. This book selected cases to analyze discourse in the busy airspace of HAF and commercial aviation in Greece, an environment of highly skillful knowledgeable professionals.

References

Adams, M., Tenney, Y., & Pew, R. W. (1995). Situation awareness and the cognitive management of complex systems. *Human Factors, 37*, 85–104.

Anosis SA (Prd). (2000). *Silence in the air (Aerines Siopes)* [Television Series]. Athen: Mega.

AOTS. (1985). *Statistical methods for quality improvement*. Tokyo: Association for Overseas Technical Scholarship (AOTS), Chosakai, Ltd.

Ashman, A., & Telfer, R. (1983). Personality profiles of pilots. *Aviation, Space and Environmental Medicine, 54*(10), 940–943.

Awbrey, J., & Awbrey, S. (1995). Interpretation as action: The risk of inquiry. *Inquiry: Critical Thinking Across the Disciplines, 15*, 40–52.

Charmaz, K. (2006). *Constructing grounded theory: A practical guide through qualitative analysis*. Cambridge/London: Sage.

Charmaz, K. (2009). Shifting the grounds: Constructivist grounded theory methods. In J. M. Morse, P. N. Stern, J. Corbin, B. Bowers, K. Charmaz, & A. E. Clarke (Eds.), *Developing grounded theory: The second generation*. Walnut Creek: Left Coast Press.

Christiansen, Ó. (2011). Rethinking "quality" by classic grounded theory. *International Journal of Quality and Service Sciences, 3*(2), 199–210.

Connell, L. (1995). Pilot and controller communications issues. In B. G. Kanki & O. V. Prinzo (Eds.), *Proceedings of the Methods & Metrics of Voice Communication Workshop*. Washington, DC: FAA.

Cookson, S. (2009). Zagreb and Tenerife: Airline accidents involving linguistic factors. *Australian Review of Applied Linguistics, 32*(3), 22.1–22.14.

Corbin, J., & Strauss, A. L. (1990). Grounded theory research: Procedures, canons, and evaluative criteria. *Qualitative Sociology, 13*(1), 3–21.

Cozzolino, A. (2012). *Humanitarian logistics: Cross-sector cooperation in disaster relief management*. Newyork: Springer Briefs in Business.

Dalrymple, M. A., & Schiflett, S. G. (1997). *Measuring situational awareness of AWACS weapons directors. Situational awareness in the tactical air environment: Augmented Proceedings of the Naval Air Warfare Center's First Annual Symposium* (CSERIAC SOAR Report# 97-01). Ohio: WP-AFB.

Dekker, S. W. A. (2010). Pilots, controllers and mechanics on trial: Cases, concerns and countermeasures. *International Journal of Applied Aviation Studies, 10*(1), 31–50.
Department of the Navy (DoN). (1993, September). *Systems approach to process improvement (Instructor guide)* (pp. 5-15–5-27). San Diego, CA: OUSN Total Quality Leadership Office and Navy Personnel Research and Development Center.
Dismukes, R. K., Berman, B. A., & Loukopoulos, L. D. (2007). *The limits of expertise: Rethinking pilot error and the causes of airline accidents*. London: Ashgate Publishing.
Endsley, M. R. (1988). Design and evaluation for situation awareness enhancement. In *Proceedings of the Human Factors Society 32nd Annual Meeting* (Vol. 1, pp. 97–101). Santa Monica: Human Factors Society.
Endsley, M. R. (1994). Situational awareness in dynamic human decision making: Measurement. In R. D. Gilson, D. J. Garland, & J. M. Koonce (Eds.), *Situational awareness in complex systems* (pp. 79–97). Daytona Beach: Embry Riddle Aeronautical University Press.
Endsley, M. R. (1995). Measurement of situation awareness in dynamic systems. *Human Factors, 37*, 65–84.
Endsley, M., & Jones, D. (2012). *Designing for situation awareness: An approach to user-centered design* (2nd ed.). London: Taylor and Francis.
Flach, J. M. (1995). Situation awareness: Proceed with caution. *Human Factors, 37*, 149–157.
Foucault, M. (1972/1995). *The archeology of knowledge* (R. Sheridan, Ed., Trans.). London: Routledge.
Freeman, J. T. & Cohen, M. S. (1995). *Methods for training cognitive skills in battlefield situation assessment* (Technical Report 95-2). Fort Leavenworth, Kansas: US Army Research Institute.
Garland, D. J., Phillips, J. N., Tilden, D. S., & Wise, J. A. (1991). *Theoretical underpinnings of situation awareness: Towards an objective measure* (Tech. Rep. No. CAAR -15498-91-1). Daytona Beach: Embry-Riddle Aeronautical University, Center for Aviation/Aerospace Research.
Gibson, J. J. (1979). *The ecological approach to visual perception*. New York: Houghton Mifflin.
Gilbert, G. N., & Mulkay, M. J. (1984). *Opening pandora's box: A sociological analysis of scientists' discourse*. Cambridge: Cambridge University Press.
Glaser, B. (1992). *Emergence vs. forcing: Basics of grounded theory analysis*. Mill Valley: Sociology Press.
Glaser, B. G., & Strauss, A. L. (1967). *The discovery of grounded theory: Strategies for qualitative research*. New York: Aldine de Gruyter.
Grice, H. P. (1989). *Studies in the way of words*. Cambridge, MA: Harvard University Press.
Grover, S. L., Hasel, M. C., Manville, C., & Serrano-Archimi, C. (2014). Follower reactions to leader trust violations: A grounded theory of violation types, likelihood of recovery, and recovery process. *European Management Journal, 32*(5), 689–702.
Guba, E., & Lincoln, Y. (1994). Competing paradigms in qualitative research. In N. K. Denzin & T. S. Lincoln (Eds.), *Handbook of qualitative research*. Thousand Oaks: Sage.
Hall, S. (1997). Introduction. In S. Hall (Ed.), *Representation: Cultural representations and signifying practices* (pp. 1–12). London: Sage.
Hallberg, R.-M. L. (2006). The "core category" of grounded theory: Making constant comparisons. *International Journal of Qualitative Studies on Health and Well-being, 1*, 141–148.
Hellenic Republic Ministry of Transport and Communications (HRMTC). (1998). *Accident investigation Report for AeroSvit AirlinesYakovlev-42 Flight at Pieria Mountains on December 17, 1997*. Athens: Ministry of Transportation & Communications/Aircraft Accidents Inquiry Council.
Hildebrand, B. (2004). Gemeinsames Ziel, verschiedene Wege: Grounded theory und Objektive Hermeneutik im Vergleich. *Sozialer Sinn, 4*(2), 177–194.
Hone, G. N. (2003, November). *The myth of SA*. Presented at the Symposium: NEC – The Human Dimension, The Defence Academy of the UK, Shrivenham.
Hone, G. N. (2008). Awareness introductory remarks (Guest editorial). *International C2 Journal, 2*(1). http://www.dodccrp.org/html4/journal_v2n1_00.html

Hone, G. N., Macleod, I. S., & Smith, S. J. (2005). Awareness and scenario-based requirements. In *Proceedings of the Human Systems Integration Symposium*. Washington, DC: American Society of Naval Engineers. Published on CD-ROM.

Hone, G., Martin, L., & Ayres, R. (2006). Awareness-does the acronym "SA" still have any value? In *Proceedings of the 11th International Command and Control Research and Technology Symposium: Coalition Command and Control in the Networked Era, At Cambridge UK*. Washington, DC: The DoD-CCRP (on CD-ROM).

Hymes, D. H. (1972). On communicative competence. In J. B. Pride & J. Holmes (Eds.), *Sociolinguistics* (pp. 269–293). Baltimore: Penguin Education, Penguin Books Ltd.

ICAO (International Civil Aviation Organization). (2001). *Aircraft accident and incident investigation/Annex 13 to the convention on international civil Aviation* (9th ed.). Montréal: ICAO (Annex 13/Aircraft Accident and Incident Investigation. ICAO Accident/Incident Reporting Manual -Doc 9156). Online. Available HTTP: http://www.emsa.europa.eu/retro/Docs/marine_casualties/annex_13.pdf. Accessed 3 Nov 2013.

Ishikawa, K. (1968). *Guide to quality control*. Tokyo: JUSE.

Jantunen, S., & Gause, D. C. (2014). Using a ground theory approach for exploring software product management challenges. *The Journal of Systems and Software, 95*(2014), 32–51.

Kelle, U. (2005). "Emergence" vs. "Forcing" of empirical data? A crucial problem of "Grounded theory" Reconsidered [52 paragraphs]. *Forum Qualitative Sozialforschung/Forum: Qualitative Social Research, 6*(2), Art. 27e. http://nbn-resolving.de/urn:nbn:de:0114-fqs0502275

Kochan, J. (2009, December 10). HCEP lecture: 'In search of the expert pilot- A research Odyssey in human factors', Drexel Human Cognition Enhancement Program (HCEP at http://www.drexel.edu/research/humancognition/). Online Video File at: http://media.irt.drexel.edu/mediasite/Viewer/?peid=8f7355a186f947798bd6bf7db224fefe. Accessed 6 May 2010.

Konstantaras, D. (2007). *24 seconds (24 Defterolepta)*. Athens: Kastaniotis Publications.

Krock, L. (2006). The final eight minutes of the deadliest plane crash. NOVA Public Broadcasting Service. PBS Online at http://www.pbs.org/wgbh/nova/space/final-eight-minutes.html (Posted 10.17.06).

Leveson, N. (2012). *Engineering a safer world: Applying systems thinking to safety*. Cambridge, MA: MIT Press.

Locke, K. (2001). *Grounded theory in management research*. London: Sage.

Loukopoulos, L. D., Dismukes, R. K., & Barshi, I. (2009). *The multitasking myth: Handling complexity in real-world operations*. Farnham: Ashgate.

Maddux, C. D., & Johnson, D. L. (2009). Information technology in education: Some reasons for optimism. *Computers in the Schools, 26*(2), 83–88.

Martin, P. Y., & Turner, B. A. (1986). Grounded theory and organizational research. *Journal of Applied Behavioral Science, 22*(2), 141–157.

Mavor, A. S., Kidd, J. S., & Prince, C. S. (1995). *Tactical display for soldiers*. Washington, DC: National Academy Press.

Miles, M., & Huberman, M. (1994). *Qualitative data analysis: A sourcebook of new methods* (2nd ed.). Newbury Park, CA: Sage.

Muller, M. (2014). Curiosity, creativity, and surprise as analytic tools: Grounded theory method. In J. S. Olson & W. A. Kellogg (Eds.), *Ways of knowing in HCI* (pp. 25–48). New York: Springer.

Muller, M. J. & Kogan, S. (2010). *Grounded theory method in HCI and CSCW*. IBM Watson Research in Cambridge. Report: 10-09.

National Transportation Safety Board (NTSB). (2010). *Loss of thrust in both engines after encountering a flock of birds and subsequent ditching on the Hudson River, US Airways Flight 1549, Airbus A320-214, N106US Weehawken, New Jersey, January 15, 2009*. Aircraft Accident Report NTSB/AAR-10/03. Washington, DC.

Nelson, L. M. (1998). Collaborative problem solving. In C. M. Reigeluth (Ed.), *Instructional design theories and models: A new paradigm of instructional theory* (Vol. II, pp. 241–267). Hillsdale: Lawrence Erlbaum Association.

Newby, T. J., Stepich, A. D., Lehman, J. D., & Russell, J. D. (1996). *Instructional technology for teaching and learning*. Englewood Cliffs: Prentice Hall.
Paltridge, B. (2006). *Discourse analysis*. London: Continuum International Publishing.
Patton, M. Q. (2002). *Qualitative research and evaluation methods* (3rd ed.). Thousand Oaks: Sage.
Peirce, C. S. (1931–1958). Collected papers of Charles Sanders Peirce (C. Hartshorne & P. Weiss, Eds.). Cambridge, MA: Harvard University Press.
Potter, J., & Wetherell, M. (1987). *Discourse and social psychology: Beyond attitudes and behaviour*. London: Sage Publications.
Reichertz, J. (2007). Abduction: The logic of discovery of grounded theory. In A. Bryant & K. Charmaz (Eds.), *The sage handbook of grounded theory* (pp. 214–228). Thousand Oaks: Sage.
Reinhardt, R. F. (1970). The outstanding jet pilot. *American Journal of Psychiatry, 127*(6), 732–736.
Retzlaff, P. D., & Gibertini, M. (1987). Air force pilot personality: Hard data on the right stuff. *Multivariate Behavioural Research, 22*, 383–389.
Ritchie, J., & Lewis, J. (2004). *Qualitative research practice: A guide for social science students and researchers*. London: Sage Publishers.
Robson, C. (2002). *Real world research: A resource for social scientists and practitioners-researchers* (2nd ed.). Malden: Blackwell publishing.
Roitsch, P. A., Babcock, G. L., & Edmunds, W. W. (1978). *Human factors report on the Tenerife accident*. Washington, DC: Air Line Pilots Association.
Searle, J. (1979). *Expression and meaning: Studies in the theory of speech acts*. New York: Cambridge University Press.
Shaw, R. L. (1985). *Fighter combat: Tactics and maneuvering*. Annapolis: United States Naval Institute.
Smith, P. L., & Ragan, T. J. (1993). *Instructional design*. New York: Macmillan Publishing Company.
Strauss, A. L. (1987). *Qualitative analysis for social scientists*. Cambridge/New York: Cambridge University Press.
Strauss, A. L., & Corbin, J. (1990). *Basics of qualitative research. Grounded theory procedures and techniques*. Newbury Park: Sage.
Suddaby, R. (2006). From the editors: What grounded theory is not. *Academy of Management Journal, 49*(4), 633–642.
Tague, R. N. (2004). *The quality toolbox* (2nd ed., pp. 247–249). Milwaukie: American Society for Quality, ASQ Press.
Tsolakis, A. D. (2013). *Dare! A way of life (Tolma! Enas Tropos Zois)*. Athens: 11Aviation Publications.
Turner, D. (2010). Orally-based information. *Journal of Documentation, 66*(3), 370–383.
Turner, D., & Warren, A. (2010, June 21–24). *Investigating oral information*. Paper presented at the Conceptions of Library and Information Science Conference, London, UK.
U.S. Air Force (n.d.). Process Improvement Guide - Total Quality Tools for Teams and Individuals, p. 33. Air Force Electronic Systems Center, Air Force Materiel Command.
Wagner, S. M., Lukassen, P., & Mahlendorf, M. (2010). Misused and missed use – Grounded theory and objective hermeneutics as methods for research in industrial marketing. *Industrial Marketing Management, 39*(1), 5–15.
Woods, D. D. (1988). Coping with complexity: The psychology of human behavior in complex systems. In L. P. Goodstein, H. B. Andersen, & S. E. Olsen (Eds.), *Tasks, eros and mental models*. London: Taylor & Francis.
Wooffitt, R. (2005). *Conversation analysis and discourse analysis: A comparative and critical introduction*. London: Sage.

Part III
Knowledge Operators of Silence and Voice

Chapter 5
Knowledgeable Sounds of Silence or When Silence Is Not Golden

5.1 Organizing the Chapter of Silence

Literature for this book primarily covers knowledge elements from systems theory and human factors, whereas material and references more closely connected with communication and culture have been worked into the main data chapters (Chaps. 5 and 6). Communication-related elements infuse new constructs and inputs in knowledge theory and remain with closer proximity to the empirical evidence described.

The conceptualization of silence underlies several aspects of how the silence concept is instantiated in the reality of the participants in this study, with seemingly related observations and experiences. Different interlocutors, in this study, perceive silence differently in their roles and connect the phenomenon with diverse practices, regulations, incidents, and concepts. Hence, the current chapter follows the different points of view of pilots and air traffic controllers, and the chapter sections reflect the dimensions that materialize silence when responders exercise their roles. That is why this Chap. 5 develops with one main scenario, whereas Chap. 6 (for voice) incorporates three distinct scenarios and a different structure.

Silence is manifested as a framing mechanism of the landmark flights explained in Chap. 3 (Sects. 3.1.1 and 3.1.2) and is taking several forms depending on the roles of aviation actors who interact. Air traffic controllers treat it differently compared to cockpit crew members and fighter pilots; cabin crew members think about it as a restricted channel; cockpit crew members experience the silence of authority. From a regulatory point of view, "sterilization" is imposed as silence in the cockpit, in certain intervals. Lastly, the lack of communication is perceived differently depending on who initiates (and then is expected to break) silence.

Silence intervals are part of the whole interaction (in Sect. 5.2), but could have devastating consequences as found in intra-crew and inter-crew research, and structure a main scenario as a communication factor in the concept of aviation safety (Sects. 1.3.1 and 5.3). After considering the scenario of "pilot silence" (introduced

T. Katerinakis, *The Social Construction of Knowledge in Mission-Critical Environments*, Innovation, Technology, and Knowledge Management,
https://doi.org/10.1007/978-3-319-91014-7_5

Table 5.1 Selected responders for the scenario of "silence" narrative

1.	Eleftherios Venizelos, Fighter Pilot-AIP1	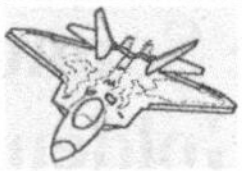
2.	Ioannis Daskalogiannis, Fighter Pilot-AIP2	
3.	Ippokratis Koos, Helicopter Pilot-AIP6	
4.	Ikaria Ikaros, ATC-AIP14	
5.	Oia Santorinis, ATC-AIP15	

in Sect. 3.2.1), the air traffic controller's point of view is analyzed (Sect. 5.4) with relevant concepts: initiating, violating clearance, pilot apologizing, feeling the hidden corporate pressures, dealing with emergency requirements, and no-talking with the collaborative scheme of pilots and controllers. Then the fighter pilot's point of view is presented (in Sect. 5.5.1) with the *instructor knows better* mentality, the mutual silence, the mentoring outcome, the safety and trust culture, its connection with situation awareness and leadership, its impact in checklist monitoring, VIP transfers, and the arbitrary and nonstandard choice of not sharing crucial information. These sections coupled with a conclusion summary are those to follow.

For this chapter I selected the responders whose interpretations went beyond a procedural predictable reply and offered insightful experiences. From the complete list of Sect. 2.5 (Fig. 2.3), the subset of five respondents (in standardized names) in this chapter is shown below:

As we can see in Table 5.1, seven participants with different profiles and backgrounds (military and civil aviation pilots, controllers, and an accident investigator) were selected for the content and multidimensionality of their responses. With their interpretations we move from the conceptualization of the scenario to the operational assessment of silence, as they experience it while doing their job. More details for these responders are included in the Appendix E.

5.2 Aviation Silence in Flight Continuum and Beyond

This section sets the background of silence in aviation communication as collected by the informal discussions with the responders of this study, prior to their commenting on the specific scenarios. The systems approach used in this book describes flight cockpits with their information exchange as mission critical operating environments. Even when mission critical systems fail, it is much more likely due to an accumulation of (in other contexts) rather minor malfunctions and trivial difficulties, as major technical or competence issues are taken for granted in complex systems[1] (e.g., Gaba et al. 2003; Barrett 1982). Several tragic accidents illustrate that when pilots or crew members fail to speak up and transmit safety-critical information, the risk for life increases rapidly (Roitsch et al. 1978; Cookson 2009).

Sometimes voicing opinion regarding elements from flight incidents (from blinking switches to checklist negligence or repetitive no-talk) has a postflight follow-up, in the debriefing phase when the pilot (captain) fills out the flight logbook[2] and explains important phenomena observed during the flight (Konstantaras 2007). Accountability goes hand in hand with breaking silence, in this occasion, although it might be related to "whistle-blower (corporate) culture" or lack of it (especially in civil aviation).

In this chapter, the discussions with high-level practitioners point out a genre of silence; the flight communication model is dominated with communication issues concerning the channel of silence. Diverse studies (Cushing 1994; Barshi 1997; Loukopoulos et al. 2001; Dismukes et al. 2007; Gladwell 2011) indicate that miscommunication is a pervasive problem in flight operations and has been a decisive factor in numerous incidents and fatal accidents, although of a low frequency when considered as a percentage of daily interactional transactions.

Pilots and Air Traffic Controllers Communicate Until and when the controller-pilot data link communication becomes widespread, air traffic control communication depends primarily upon voice communication (and the periods of silence which communicates too) and is affected by various parameters. Controller-pilot data link communications (CPDLC) system aims to reduce the number of voice messages from air traffic controller by sending electronic messages that are displayed on a computer screen in the cockpit, using a keyboard of *words* (and not just letters).[3]

---

[1] Beyond the aviation industry, miscommunication issues are prominent in other mission critical environments like medical practices. Also, lack of a regularized communications system with predefined terminology was identified as a contributing factor in the nuclear accident at the Three Mile Island Nuclear Power Plant operated by Exelon, Dauphin County, PA, in 1979.

[2] The aircraft cabin defect log is usually filled by the cabin crew members and the aircraft technical log is filled in by the flight deck members. The two logs combined communicate flight realities to the next shifts and support personnel, whereas the two flight recorders – CVR for voice and FDR for instrument indications and parameters – are the real-time archive of the flight.

[3] The controller-pilot data link communications for FAA is explained at http://hf.tc.faa.gov/capabilities/cpdlc.htm. In Canada, airspace has undergone safety and efficiency enhancements with the completion of the national implementation of CPDLC in suitably equipped aircraft flying above 29,000 feet (http://www.navcanada.ca/EN/media/Pages/news-releases-2014-nr10.aspx). Eurocontrol (at http://www.eurocontrol.int/services/controller-pilot-data-link-communications) is using it on a simulation basis. The Australian Civil Aviation Safety Authority (http://www.icao.int/APAC/Meetings/2013_FIT_Asia2_RASMAG18/IP04%20Use%20of%20CPDLC%20%20ADS-C%20in%20Australia.pdf) is evaluating that technology after 15 years of use in the oceanic air space.

The responsibilities of the pilot and the controller overlap in many flight areas and provide backup and complementarity, as both parties are equally involved from different edges of the flight. Failure of speaking up (or reading back) or replying on time, ambiguous terminology, and/or the omissions of key words or phrases might result to information misunderstandings, to not knowing the *here and now* condition. When single radio contacts are used for the oral transmission of essential information, then vulnerability to potential dangers is apparent (Job 1994).[4]

There are several critical phases in cockpit-air traffic controller communication cycle or continuum (as analyzed in Sect. 1.3). Silence is part of the interaction, and voice should be used to formulate the signals of interaction (till the computerized data link will become widespread). The phases of a flight, in civil aviation (and not only), include the briefing on the ground (for the parameters of the flight like network of towers en route, navigation altitudes, weather conditions, fuel load, navigation aids, expected baggage load, etc.). Then, the preflight ritual of walking around the aircraft for a visual check[5] and the studying of aircraft flight logs (for potential unusual situations documented in the log, as it happened in the Falcon VIP flight of Sect. 3.1.1). Then, the marking of checklists in the cockpit (communicating with ground control and crew) and door closing and cross checking. Then, getting clearance and starting the takeoff roll and landing gear up.

Then, communicating the banner of "seat belt *sign*" off (except during oscillations); flying to 20,000 ft. (on average) and continue climbing. Then engaging the autopilot (from manual to automated control) and communicating with departing-intermediate-destination control towers (the "Pilot Not-Flying" as "Pilot Monitoring" monitors the flight). Automatic recording of flight indicators and intra-crew interaction are recorded in CVR and FDR.

When approaching the destination, topical awareness (like LA of Sect. 4.3.2) is exercised with interaction between co-pilot and air traffic controllers. By disengaging the autopilot (from autopilot to manual control), the "Pilot Flying" continues with landing gear down. Then, by obtaining clearance to land from the

[4] Nine months after the fatal collision of two Boeing 747 s at Tenerife's accident on 1977, with 583 fatalities, the ICAO Air Navigation Committee took action, issuing 3 reports and implementing radiotelephony changes in 1984. Among the contributing factors, the expression *at takeoff* determining a position different than expected created a major misunderstanding. Twenty years later and further on, miscommunication still occurs and remains a contributing factor to aviation accidents.

[5] In October 2, 1996, the Aeroperu Flight 603 with a Boeing 757 from Lima (Peru) to Santiago (Chile), the maintenance personnel forgot to remove the protective covers of the static ports (the external pressure-sensitive instruments used to sense altitude, wind speed, temperature, etc.) after washing the fuselage of the aircraft. As a result, during the flight, essential indicators were measuring erroneous inputs that leaded to erroneous understanding of the flight parameters. It seems that the oversight could be prevented with the common (for crew members) preflight walk-around the aircraft. The crew did not walk-around that aircraft, and having inadequate situation awareness in the flight from faulty indicators (as well as problems in navigation, standard call-outs, and tactical decisions), they did not prevent the fatal crash in the Pacific Ocean (with the altimeter erroneous indication of 9,700 feet altitude in the impact to the water), and 70 people on board lost their lives. The accident report from the Directorate General of Air Transport (of the Ministry of Transport, Communications, Housing and Construction) of Peru is available at Eurocontrol's skybrary at http://www.skybrary.aero/bookshelf/books/1719.pdf.

controllers and landing the aircraft, stopping or turning off active runway and arriving at terminal (with constant interactions with controllers), the flight meets its goal. A debriefing phase is conducted if needed, as well as completion of flight logs. Silence periods are required after the landing gear up and till the "seat belt sign off" and from the landing gear down till landing (as explained in Sect. 2.1).

Military Flights In the case of military flights, a m-ission is allocated, configuration and flight planning is done on the ground, briefing is a crucial phase as tactical and communication details are posed, and then the pilot is doing a visual check of the aircraft, climbs on board the fighter plane, is going over a checklist with the ground engineer, gets clearance to take off, and starts the mission. During the mission there is intense communication with the accompanying fighter planes and the Center of Operations.

These interactions are recorded in an FDR device, silence is imposed when mission target is identified till the end of the fight or drill, and then the fighter pilots approach back to the base and get clearance and landing gear down to land. The first to communicate is the ground engineer who is assigned to the fighter plane and gets all the necessary remarks from the pilot. Debriefing is the last but very important phase (on the ground), when all preflight arrangements are compared with the actual flight, as well as the communication interactions to evaluate an accomplished mission or erroneous situations and prevent unexpected events (in future flights).

Silence sometimes occurs from landing to debriefing, as fighter pilots are inclined to share opinions when they are between each other. All fighter pilots who participated in this study underlined (briefing and) debriefing as an integral part of any mission in the (preflight and) postflight phase. The debriefing is an account of what happened in the flight and of "who needs to know what" in order to fix, arrange, and prevent any occurrence. The debriefing is a meta-communication process which retracts the whole flight, provides a mechanism to evaluate human error and safety culture in a systemic way, and cultivates collegiality and leadership.

These series of stages cover an approximation of a routine flight; the ideal flight is described in the flight manuals; implementing it while airborne is the actual flight; flying with no unresolved occurrences is the routine flight; and dealing with major events or conditions leading accidents is also part of the real flight situation (which includes all but the "ideal" flight, in the sense explained in Sects. 2.1 and 2.3). So, periods of silence are part of the communication process, but the duration of these periods, the way silence breaks into messages (vocal or protocol), and the consequences of different manifestations of silence are the objective of this chapter.

5.3 Knowledge for Silence, Addressing Safety

In the following sections, the silence scenario presented in Sect. 3.2.1 is used for the discussion with the informants-interviewees, who offer their commentary in writing. The point of view of different actors (pilots, controllers, crew members, accident investigators) determines various manifestations of silence and events in which silence was decisive for the conclusion of the flight or the safety of passengers on board.

Since the early years of (mainly) civil aviation, decisions whether and how crew members voice concerns involve complex considerations and trade-offs. Reluctance to speak up and the resultant silence have posed a serious threat to flight safety, and many discussants comment on the level of flight risk which justifies the cost of speaking up. Barriers for voicing concerns could be relationship damage and contextual factors (such as familiarity with the people in the alarming situation). Silence, in the interpretation of the responders for this study, could be observed inside the airplane cabin, inside the airplane cockpit, and between cockpits or may be assessed institutionally with the flight log reports where the crew must fill any noteworthy information (for the airline and the aircraft manufacturers).

The scenario, I use for silence, originates from the research by Bienefeld and Grote (2012) who report that several accidents have shown that failure to speak up can have devastating consequences (the full narrative is in Sect. 3.2.1). In this book, silence acquires multiple dimensions which clearly extend the original position on the "silence" theme. Despite the durable crew resource management (CRM) training, silence persists as a problem and still poses a risk to flight safety. In high-risk situations (like flights), speaking up plays a pivotal role in conveying important information to mitigate errors.

Silence might occur in intra-team and inter-team communication and in ad hoc cases, where the perception of "feeling safe to speak" mediates the relationship between status and speaking up and between perceived leader inclusiveness and speaking up (Grote 2014). Leadership inclusiveness describes the words and acts (or lack of action) that leaders undertake to invite and appreciate other's contributions and accommodate other's choice to speak up (Nembhard and Edmondson 2006). As shown in the responses of the current chapter, a leader could be a pilot-in-command (for example) who leads the crew and directs the interaction with controllers. A leader (as supervisor) could be the commanding officer in a fighter squadron or the manager in an airliner. The silence behavior of subordinates in organizational settings is closely dependent on the supervisor's attitudes towards silence (Vakola and Bouradas 2005). Supervisors create the microclimate (a certain vocational subculture) to motivate subordinates to defend safety rules or to remain silent.

Silence and voice could also relate with the uncertainty of interaction, which in turn relates to safety. Grote (2014) argues that in certain circumstances of high-risk environments, increasing uncertainty could forward. Not remaining silent empowers people to address concerns and doubts about certain actions for sounder decision-making, even in conditions of added uncertainty. In high-risk situations, flexible rule-making with degrees of freedom for the actors could raise uncertainty for the decision-maker but could be more appropriate and safety-increasing. The priority on safety of passengers and the aircraft may be a catalyst for both, voice and silence.

During the study for this book, the responders use the narrative of the silence scenario to offer their opinion about the crew members and to explain a similar experience from the controller's point of view and as a pilot who is expected to speak up. The probing questions to elaborate initial responses concern the likelihood of a fighter pilot mission supervisor (who assists the controller in military missions) for coordinating silence and dealing with authority (of rank or task at hand), as well as the potential training required. In fact, silence and voice could be problem-focused, suggestion-focused, and opinion-focused in different situations (Schwappach and Gehring 2014). Also, responders are asked to take the point of view of their role (and then add their personality), so that silence would emerge as a phenomenon with multiple viewpoints: silence from whom, addressed to whom, used by the controller or the crew member, perceived by the controller, or a crew member, an external (intercepting) pilot, or the airline operator.

5.4 Pilot Silences from the ATC Point of View

5.4.1 Pilot Action, Latent Pressures, Emergency, and ATC

One of the most experienced air traffic controllers-informants in this study, namely *Ikaria Ikaros*, is speaking her mind for this scenario. Ikaria Ikaros is a well-respected controller working in a chief position in one of the busiest airports in Greece. While we had several face-to-face and tele-discussions in the past 3 years, it was clear to me that she understands aviation in its entirety as a "holon," i.e., aviation system and subsystems, multiple actors and public policies, social conventions and operational hierarchy and decision-making, as well as the information flow. The term "holon" (Ostrom 2005, as reflected from Koestler 1973) is used to facilitate analysis of social hierarchy and human decision-making in sub-wholes which portray behavior governed by rules or have stable properties. Sometimes the parts may not be descriptive of the holon which they have created: the operating cockpit is something beyond the people, machines, and processes that create it. In-flight communication flow and exchange may be more representative of the "holon cockpit" with the controller's interaction included.

All the other controllers respect Ikaria Ikaros as a knowledgeable authority and were prompting me to ask her for any challenging question, as well for the policies regarding the unification of European airspace (the Single European Sky initiative[6]).

[6]Launched by the European Commission in 1999, it is the initiative of organizing airspace into functional blocks, according to traffic flows rather than to national borders. The European air traffic management (ATM) system currently handles around 26,000 flights daily, with a projection to double by 2020. With common rules and procedures at European level, its primary aim is to meet future capacity and safety needs through legislation. The Single European Sky second package (SES II) established goals in key areas of safety, network capacity, effectiveness, and environmental impact. The Single European Sky transforms the role of EUROCONTROL, which could become the Network Manager of the European ATM network (description available at https://www.eurocontrol.int/dossiers/single-european-sky).

Explicit Declaration of the Situation Let us see what she had to say:

> It happened throughout my long carrier as an air traffic controller to experience emergency situations or other situations needed to be handled according to the rules, after a pilot's report, declaring a situation. In my opinion, when there is an imminent danger for the aircraft, a real emergency situation (i.e. high jacking, fire, fuel loss, engine failure), the crew member communicates this to ATC. In some other cases, where only indications in the normal operations of the flight were declared, the crew member was hesitant to speak up about the nature of the problem, in order the ATC staff to be prepared accordingly, in case the situation would turn worse.

Ikaria Ikaros states that the controller is trying to be prepared for the current and next projected flight inquiry (with TA for GA) when explicit declaration of a situation is coming from the pilot, preferably in low-context language. Communication is initiated by the pilot side and may include an actual textbook emergency or indicators that may accumulate. The controller has more response resources on the ground but needs a clear assessment from the cockpit. Simultaneously the controller tries to be economical in effort and conversation, as multiple possibilities may be expected in the coming future for each flight. Sometimes, when the situation evolves, and the crew member sends a hesitation message with some uncertainty, the controller needs to stay vigilant for a real emergency in terms of flight control, fire, and fuel or engine loss. And why does she think that happens?

> I suppose that beside the stress occurring from workload in the cockpit there is always a "hidden" pressure related to the so called "fame of the company", in other words, the crew members hesitate to frankly declare the nature of the problem, just leaving the controller to suggest what is wrong (in those cases we are always in a local standby phase). If everything ends up well, we don't get any further information, than simple phrases like "just a small problem, nothing serious, only a small indication, etc."

Ikaria Ikaros values the strain crew members feel from their navigation workload and the additional corporate latent practice to preserve the airliner's fame. If communication is not initiated by the crew to speak about the problem, then the controller starts prompting questions in order to suggest what is wrong. It is an unorthodox practice as the controller remains occupied in a *waiting to continue* mode. When the problem resolves, a short message about *a small nonserious indication* is the crew follow-up response.

I asked Ikaria Ikaros to explain the implications further, and she is clear that it is not the controller's job to speculate for a flight problem, but it is an asymmetrical shift of responsibility; each part (pilots and controllers) controls its own territory, manages its own topical situation, and should provide input/feedback to the other about the next step of the flight from its own oversight. Then they can start symmetrical interaction to collaborate. That is why the one who has a potential problematic situation declares it, and communication continues from there.

The cockpit is a stressful environment since human actors need to manage sophisticated devices, interact and communicate, adhere to tight time frames, and implement a flight that carries the responsibility of passenger lives and the aircraft itself. When indicators of a potential problematic situation are evident, the stress factor increases. As cockpit members are the "spokespersons" of the airliner to the air traffic controllers who represent the aviation authorities, their interaction has an impact to the airliner brand, as competition in aviation is not only about customers but also means to maintain strong perceptions in the minds of the aviation community members.

Aviation is a highly regulated environment[7]; any problem report is an operational disruption when the crew cannot deal with that internally. So, giving hints instead of explicit language is a sort of transferring the load to your responder, who is the controller in these cases. This hesitation to speak frankly has the form of thinking aloud that may be forgotten if everything goes well. Of course, there are cases when the end of a flight is the start of a follow-up process to report minor problematic issues, as in the example analyzed for the Falcon VIP flight (in Sect. 3.1.1).

Report-Encouragement Practice and Silence But why is this important, as Ikaria Ikaros continues?

> This can be proven very dangerous behavior as there is always a possibility for an indication to turn into a disaster. I do not know about airlines' crew training policy regarding this issue, but as it happens quite often with different companies, I believe that there is rather a fear among crew members to speak up in such cases without having in mind the consequences of this decision (to speak up freely about the "problem").

Ikaria Ikaros explains that the report-encouragement policy goes hand-by-hand with vigilance for the future. Bits and pieces of earning signals, communication, and mechanical indicators can lead to disasters. Cloudy information from the flight deck and small-talk-oriented conversation may cause inefficient interaction and insufficient information. Also, a process of outeraction develops when participants (cockpit crew to controllers) step outside information exchange interaction and negotiate availability to speak to each other (even in another secondary frequency) and establish an attentional contract which binds them tightly together (as described in Nardi et al. 2000; Hwang et al. 2009). In this sense they find a way to manage their communication (in case needed, out of the routine) and not so much to exchange information.

The consequences of these interactions (and corollary reports) may be long term if the airliner support and facilities chain do not hit a specific verbal comment to follow. Fear to speak up could also be a symptom of loose or repressive organizational culture at an airliner level, with an ill-perception of human-error blame (see Dekker 2011), or microculture developed inside the cockpit and the cabin. So, as Ikaria Ikaros presents her reasoning about *the crew not-being silent* argument, she is conscious on the regulated counterargument when, sometimes, silence is required. She prompts silence as a "magic" word (in capitals to accentuate the requirement) below:

---

[7] IATA's director general Tony Tyler is quoted saying: "Regulators are micro-managing our businesses, telling us how we may advertise our services, how long we must hold a reservation that has not been paid for and how we are to manage operational disruptions regardless of the cause. These regulations impose a huge penalty on the economy and ultimately raise the cost of air travel for all consumers" (see September, 20, 2013 at http://www.aviationtoday.com/the-checklist/IATA-U-S-Needs-to-Loosen-Regulatory-Grip-on-Airline-Industry_80223.html#.U_dM6qN0Ykx).

> Last, I would like to mention that during emergencies crew members have to focus on how to navigate the aircraft, so they need time and peace. SILENCE is the magic word imposed to the Controllers in order to give time to the crew to identify and solve the problem in the sake of the safety. So, SILENCE is sometimes necessary for pilots and controllers.

Aviate, Navigate, and Communicate The rebuttal about "not being silent" is what Ikaria Ikaros is clear to say during emergencies; that having silence intervals could be an operational requirement when real emergency occurs (*Mayday* opening call) or distress (*Pan Pan* opening call). An initial call incorporating the word *standby* would be a typical feed-forward in such occasions. In emergency cases or abnormal situations when normal standards[8] may not suffice for a safe landing, as Ikaria Ikaros elaborated in further discussion, flight crews usually prioritize their actions in three steps:

Priority of Action in Flight (with a Loop)
Aviate (ensure the safe flight path and condition of the aircraft). Navigate (continue the flight to the originally intended destination, divert if necessary, carry out an emergency descent, or just place the aircraft in a safe flying position). Communicate (declare an emergency ASAP and cancel it later if the situation allows).

She is considered a fair-minded flight traffic "supervisor," as several of her trainees now colleagues agree on, and has capitalized silence as a "magic word" since controllers need to ensure that they facilitate the incoming flight with giving them *time* to manage the emergency (the controller in standby mode), free the *airspace* in their surroundings (calling the other planes to holding patterns), and impose *silence* (to the radio frequency or provide an alternative frequency). Of course, respective

[8] The Robert Zemeckis Paramount's mystery-thriller "Flight" (2012) uses one such abnormal situation scenario when an airline pilot miraculously crash lands his plane after a mid-air catastrophe, saving nearly every soul on board, despite that the investigation into the accident reveals troubling physiological condition issues of the pilot (see more at http://www.paramount.com/flight/). In Greece, as HAF is considered to be part of the everyday lifeworld of the people, a national TV series of Anosis/Errikos Anagnostopoulos called "Silence in the Air (Aerines Siopes, 2000s–2003)" is based on the personal/operations life of some pilots of the 332 Al Weather Squadron of 114CW. Some actions in the 98 3-year episode series are inside the air base. In episodes #55–65, fighter pilots act or comment on nonstandard situations (more at http://www.impdb.org/index.php?title=Aerines_siopes). Silence is featured, as causing major incident in fighter training flights, as understanding your own error (and better not to speak till the debriefing), or as feeling so comfortable with the other with whom you communicate without words. This narrative created a popular perception of the "silence" question in the broader audience. On 2016, the Warner Bros Clint Eastwood's film "Sully" narrated the "Miracle on the Hudson" when Captain Chesley Sullenberger glided his disabled plane onto the frigid waters of the Hudson River, saving the lives of all 155 aboard. However, even as Sully was being heralded by the public and the media for his unprecedented aviation skill, a NTSB investigation with simulation and manual reenactments took place to clarify the "unexpected event" and pilot's decision-making beyond any training experience (more at http://www.imdb.com/title/tt3263904/, http://www.sully-movie.com/#home).

silence intervals may be needed for the controller to activate the necessary ground services that deal with emergency response and to keep a mind in "peace" to coordinate the rest of the incoming traffic, appropriately. What otherwise would be a controller's guidance, in an emergency case, could be perceived as communication noise from the pilot who dedicates effort to aviate, navigate, and choose how to communicate (after a potential initial signal of standby). That is part of the controller's assistance.

Ikaria Ikaros, in general terms, as an experienced controller has explained silence in three dimensions: when she is the receiver of the message, when it modifies the content of cockpit communication, and when it is required in the manuals. Also, she is quite cautious to respond (as the controller with GA) to any signal initiated from a crew member reporting an actual emergency or reporting situations that may escalate to an emergency. Lastly, since civil aviation operates in a business environment with organizational and managerial regulations, crew members are employees and are subject to managerial requirements and market needs towards their passengers and the other members of the aviation community or industry; they have their corporate fame to protect and their professional face to defend.

Communication theory includes silence in the nonverbal communication mechanisms, as communication without words (Burgoon et al. 2010). Silence emphasizes the vocal channel, but with no voice, in the basic model of communication as it may relay information and influence flight interlocutors at a greater extend in compare to verbal communication. Silence serves important functions in communication (Jaworski 1997): provides time to think, may hurt others, reduces anxiety, prevents communication, and communicates emotions. It is noteworthy that when a situation is happening, unfolding or accumulating the pilot needs to explicitly declare the case (with TA evolving), interaction must start from inside the cockpit (where LA occurs), as if it was a linear process, since all indicators and warning signals are there and there is no room for interpretation and perceptual errors. The crew members have the corporeal presence in the cockpit (and the cabin for the other crew members) and are able to observe and consequently communicate factual statements (Weinberg 1958).

Another cockpit crew member would be the one who announces a minor problem in order to maintain a level of "authority" towards the controller; important situations, i.e., high jacking, fire, fuel loss, and engine failure, are more likely expected to be announced by the pilot to form a distinct message relay compared to minor indicators. If silence persists and is perceived as loss of communication, transponder switch-off, or wrong setting, then the controller is reactive to conclude a national security threat coined as "Renegade[9]" and engage interception fighters. Thus, the role of the sender-pilot is a key starter in order for the controller to apply or evaluate the next steps, when a potential severe situation is unfolding.

Ikaria Ikaros is extremely cautious in her wording "after a pilot's report, declaring a situation… the crew member communicates this to ATC," as typically, there is

[9]As explained in the second paradigmatic flight of Helios 522 "ghost plane" on Chap. 3 (Sect. 3.1.2).

no systematic way to determine whether declaring an emergency was warranted by the circumstances and it requires priority handling. In general, all traffic in the system is handled on a first-come, first-served basis. Priority handling means no undue delay and not necessarily urgency or distress; it could be a medical transport mission or search-and-rescue operations or an emergency. While, if such a case, Ikaria Ikaros is waiting from the crew member to break silence and declare the state of the flight, she wants to be able to distinguish between urgency and distress[10] and avoid the phenomenon of indiscrimination in language (DeVito 2012).

In ICAO provisions,[11] an urgency call is introduced with *Pan Pan* (expecting three repetitions), without requiring immediate assistance, and distress call is introduced with *Mayday* (expecting three repetitions) and demands immediate assistance, i.e., provide direct routing to an airport, reroute other planes to avoid delays, and not use holding for the plane that receives priority handling.

Both pilots and air traffic controllers have responsibilities for the safety of an aircraft flight and passengers, concurrently. The pilot-in-command of the flight explains his/her intensions, deviates from the normal rules and clearances, and demands maximum assistance from the air traffic controller. After speaking with a large group of controllers, their bottom line was that "in an emergency the controller can break all the rules to assist the pilot." The controller is there to assist and not just to enforce the rules when emergency is declared.

On the other hand, examples of second-guessing a pilot's decision to declare an emergency are not common, and if there is no heavy traffic for ATC, nothing will happen. If traffic was really disturbed and controllers' handling created a traffic accumulation around, combined with suspicion of an overstated declaration of emergency, then the occurrence would be subject to inquiry by the state authorities. Silence can hurt communication (Ehrenhaus 1988), as Ikaria Ikaros points out: "This can be proven very dangerous behavior as there is always a possibility an indication to turn into a disaster… just leaving the controller to suggest what is wrong." Silence should be broken from the pilot part in order to communicate an actual representation of reality through appropriate vocabulary.

Ikaria Ikaros' handling of proper feedback depends on how silence prevents communication to happen, in order to avoid *a boy who cried wolf*[12] phenomenon of false alarm or vigilance gap (Shuang et al. 2010; ADPC 2006; Roulston and Smith

[10] ICAO provisions which are explained at Flight Safety Foundation/Flight Safety Digest, August–November 2000, available at http://flightsafety.org/files/alar_bn2-3-communication.pdf.

[11] Urgency is a condition concerning the safety of an aircraft or of some person on board or within sight, but which does not require immediate assistance. Distress is a condition of being threatened by serious and/or imminent danger and of requiring immediate assistance.

[12] It refers to one of the Aesop's fables known for rhetorical storytelling and parabolic communication explaining the maxims of quality in communication: "a liar who cannot be believed even when telling the truth." When emergencies are declared repeatedly and are not real cases, then the expected response deteriorates if when the real emergency occurs. Sometimes it can be a side effect of personnel inaction after too many drills for preparedness or long-lasting readiness shifts. Remaining proactive requires "stopping for real action (an action break)" and then continuing to stay vigilant.

2004). During my duty as air traffic control monitor liaising between US Navy and Hellenic Air Force, it became customary to deal with the so-called expected emergencies, when Boeing C-17 Globemaster III[13] planes were flying to Souda Bay in Crete, after declaring emergency 3 h prior to their landing, in order to expedite diplomatic clearances for landing. After the second repetition, those emergencies became anecdotic stories for the US Naval base personnel (it was a non-emergency but was declared as one).

As Ikaria Ikaros recognizes, *hidden pressures and fame of the company* could be additional factors when silence is occurring, even in cases that crew members think they want to speak up. Aviation companies' philosophies can vary significantly and affect crew members' decision-making. Crew members may be susceptible to social compulsion in the form of peer pressure, employment pressure, and socioeconomic pressure, especially in the current era of economic turmoil. Main concerns are fears of damaging relationships and of punishment or operational pressures. In several cases, especially in the headlines of popular press, crew members opening up with problematic situations have been contrasted to a whistleblower's culture that became equivalent to transparency, although quite different by definition (AHI 2002).

A striking response in the discussion with Ikaria Ikaros was that loyalty to the company or the organization does not mean enclosure, but defining the situation in reality and engaging both crew members and air traffic controllers for a safe flight. In such cases, her opinion is that speech really acts where silence shows unacceptable inertia. It is a transition from facts (the actual situation in the cockpit) and norms (the pilot has the power and control to communicate) to act in communication as described by Habermas (1998). In this interpretation, the safety imperative of the flight requires a pilot-initiated message in terms of jurisgenesis. The pilot's talk has the communicatively generated power of a law for the controller, as both members have the rationality to reach mutual understanding. Then the controller is able to react appropriately according to the pilot's requirements.

Aviation community creates a social network of competition in market terms and solidarity social "neighborhoods" in employment terms due to the intense nature of the work, the rotating shifts and the sense of belonging of its members (Christakis and Fowler 2009; Wellman 1983, 1999). Microcultures or inner cultures are developed in between the members of aviation community covering a punitive or motivating culture of an airline or a civil aviation authority. Litigation is a practice to consider in the mind of controllers, as their focus is in cases that emergency is declared and confirmed, and their behavior as receivers is affected by a propensity to blame. A typical example is that formal reports on incidents drop down when

[13]As Boeing company describes it, the C-17 Globemaster IIIA high-wing, four-engine, T-tailed military transport aircraft, the multiservice C-17 can carry large equipment, supplies, and troops directly to small airfields in harsh terrain anywhere in the world day or night. The massive, sturdy, long-haul aircraft tackles distance, destination, and heavy, oversized payloads in unpredictable conditions. It has delivered cargo in every worldwide operation since the 1990s (http://www.boeing.com/boeing/defense-space/military/c17/).

airliners are about to public safety statistics, and anonymous pilot reports freeze in periods of accident investigations or furious media coverage. There is a cacophony of over-coverage in media outlets regarding aviation accidents which creates a culture of fear for the lay person (Glassner 1999).[14]

In Greece, where aviation is closely related to the everyday life of the people, the private spheres of controllers and pilots contribute to the broader debate of the private individuals who come together with them in the public sphere (Habermas 1991). Media coverage regulates and transforms the public sphere, as crashes for both Air Force and civil aviation challenge the airworthiness framework that the state regulates in Greece (Tsolakis 2013; Konstantaras 2007). The media operate in the mode of a follower (reporting ongoing events, disseminating information to the public) and in the mode of the leader (for the social construction of the problem); in the latter mode, media reflect back to the public opinion when after a disaster, for instance, as news wave bursts based on *magnification of a single-perspective narrative* (Vasterman et al. 2005). This way, impression constructs knowledge (sometimes falsifiable) in the mind of the lay person.

The single-perspective narrative forwards the propensity to blame. The airliner or the state is sometimes inclined to respond to the public opinion by pointing out an individual's human error instead of projecting an error in reference to the system. When individual blame stays in the forefront, there is a temporary sense of fulfillment for the public opinion, the airliner uses it as a motivational tool for silence, whereas the aviation community individuals remain in the shadow of a *sword of Damocles*[15] which echoes a public sentiment and individual fear (of partial just culture conditions for pilots or controllers). Beyond the notion of impending doom, aviation community members realize the risks associated with the profession and the media coverage of their responsibilities.

Immediately after, crew members adjust to the view of people, objects, and events in the way they are talked about. It is a case of intentional orientation in communication theory. There could be a conscious reluctance to speak or unconscious reluctance to admit a difficult situation, a tendency to underestimate what is happening, when it is uncomfortable; this is what most controllers agree when they have experience from both small and bigger airports in cases like Greece with a dense network of airports.[16]

---

[14]Aviation Safety Alliance currently consolidated in Flight Safety Foundation organized a 1-day conference on "Media Coverage of Aviation Accidents" with Barry Glassner as a keynote speaker; airline safety and the media coverage of *airline accidents* generated exaggerated fears of flying was his topic. More can be found at http://www.c-span.org/video/?156444-1/media-coverage-aviation-accidents.

[15]It is a moral parable from Dionysius, the tyrant of Syracuse at the end of the fifth century, B.C. The glamor and prestige of the aviation profession (especially for pilots) comes with silence as protection of fame, with the risk of blame, and with the actual risk of a fatal accident even when a flight is smooth till an emergency happens. As Dr. Daniel Mendelsohn, Charles Ranlett Flint, Professor of Humanities in Bard College, explains in his position in National Public Radio "All Things considered," August 19, 2011 (at http://www.npr.org/2011/08/19/139799434/sword-of-damocles-reference-sometimes-misused).

[16]Hellenic Civil Aviation Authority oversees 46 airports all over Greece for a population of less than 11 million (see more at http://www.hcaa.gr/content/index2.asp).

In Ikaria Ikaros response "...without having in mind the consequences of this decision," silence has also another intriguing source, the potential of a human error that is often a contributing factor in aviation accidents, either something is physically broken or an error occurred (Leveson 2002; Dekker 2002). That is why when flight status is reported, the theory of e-prime applies; events are described more accurately without the verb *to be* using standardized abstract nouns (Klein 1992; Wilson and Sperber 1988). So, when crew members think that a situation is evolving in the cockpit, then silence (and peace of mind) is needed, or they take silence as a requirement of the so-called sterile cockpit rule[17] that reads in the following extract (from FAR 121.542):

> No flight crewmember performs any duties during a critical phase of flight except those duties required for the safe operation of the aircraft... Activities such as eating meals, engaging in non-essential conversations within the cockpit and non-essential communications between the cabin and cockpit crews, and reading publications not related to the proper conduct of the flight are not required for the safe operation of the aircraft. Critical phases of flight include all ground operations involving taxi, takeoff and landing, and all other flight operations conducted below 10,000 feet, except cruise flight.

In several cases, silence is considered as not engaging in "nonessential conversation" and flight deck members can claim that. "Silence is the magic word imposed to the controllers in order to give time to the crew to identify and solve the problem in the sake of the safety," Ikaria Ikaros pointed out citing the "ASSIST" document[18] when an emergency was declared: acknowledge the call, separate the aircraft from other traffic to give it room to maneuver, silence on the radio frequency and provide an alternative frequency, inform those who need to know (emergency readiness units and authorities), support the pilot in any way possible, and time to give to the pilot to collect their thoughts.

ATC as the Message Sender When reversing the question, I ask Ikaria Ikaros from the other side of the communication model when she (as a controller) is a sender of messages for cases she preferred to remain silent. Her recollection, during her steps as a young controller, was to fall silent at times, due to natural fear from lack of experience. For example, evolving from domestic flights to international flights air traffic control, the matter of institutional rank and authority was an obstacle for her not to speak up. The fact that she was a young controller, a woman in a men's aviation world and her inclination to feel confident and mature in the profession were part of her thinking when she fell silent. As the shifts are quite organized in the sector, she did not meet any formal obstacle to speak, but it was more her own

---

[17] Eurocontrol's skybrary explains: the sterile cockpit/flight deck concept involves the restriction of flight crew member activity to that which is operationally essential during busy phases of flight – taxi out, takeoff, initial climb, intermediate and final approach, landing, and taxi in. FAA has the Federal Aviation Regulation FAR 121.542 (from which the above extract is taken) often referred to as the "sterile cockpit rule," and EU Commission has EU-OPS 1.085, a less explicit operating regulation, at http://www.skybrary.aero/index.php/Sterile_Flight_Deck.

[18] The acronym (**A**, acknowledge; **S**, separate; **S**, silence nonurgent calls; **I**, inform supervisors; **S**, support the crew; **T**, time) derives from the "Guidelines for Controller Training in the Handling of Unusual/Emergency Situations" at Eurocontrol's skybrary at http://www.skybrary.aero/bookshelf/books/15.pdf.

process of feeling capable to independently direct international flights and reply to unexpected questions, after the familiarity of domestic ones. Her operational maturity started becoming apparent in the pre-shift briefing in which controllers discuss the current and expected air traffic and exchange ideas on the status of their control system and any peculiarities they had to address.

Gradually, she was able to reply to questions of others and explain the constantly changing European guidelines to which they had to comply. The strong foundation on the knowledge of the legislative framework and the way to apply it in pilots' inquiries and handling several emergencies were the indicators of her entering a stage of personal safety and competence. But authority in this sector was not because of the positions she took but because of the unusual situations she had to cope with in front of her peers on duty. Since silence is not a training item per se, she emphasized the management of emergencies during the flight when silence is kept by pilot and controller for a time, as pilot has to navigate the aircraft and then to communicate. In those times, anything else would be considered significant noise in the communication model.[19] It is a realization of the maxim of quality and the principle of cooperation in communication.

5.4.2 *Suspending Talk, Violating Clearance, Repairing, Collaborating*

Let us now look at the response to the same scenario inquiring on "silence" from an air traffic controller, namely, *Oia Santorinis*, with a different profile, being the youngest of all responders. She still considers the manuals as a primary knowledge resource but values observation of more experienced controllers when she is not in shifts. During institutional shift rotation, she worked extended shifts in the Athens Area Control Center – that is the first level of approach in Hellenic Air Space and Flight Information Region (FIR)[20] – and was on duty in the record day of 3,467 aircrafts (approaches to land) in a single day for Athens Area Control Center.[21]

---

[19] Distress and urgency (as perceived by the pilot to the controller) are mentioned to ICAO ANNEX 10 VOL II – 5.3 Distress and urgency radiotelephony, 5.3.2.3 communication procedures – "Imposition of silence: The station in distress, or the station in control of distress traffic, shall be permitted to impose silence, either on all stations of the mobile service in the area or on any station which interferes with the distress traffic." See at http://www.icao.int/Meetings/anconf12/Document%20Archive/AN10_V2_cons%5B1%5D.pdf.

[20] Flight Information Region (FIR): An airspace of defined dimensions within which flight information service and alerting service are provided. A three-dimensional area, in which aircraft are under control of usually a single authority. Internally an FIR is divided into several geographical areas called sectors. Sometimes one or more FIRs have a combined upper area control. It is a Eurocontrol definition at http://www.eurocontrol.int/lexicon/lexicon/en/index.php/Flight_Information_Region.

[21] On July 28, 2014, when I last visited Athens "El. Venizelos" airport, I was able to take a note for the upcoming report. This number of flights in a single day corresponds to the (approximately) 14% of the total number of flights in the Eurocontrol Air Space of 38 countries (explained at http://www.acac.org.ma/ar/Workshop%20Presentation/Microsoft%20PowerPoint%20-%20Eurocontrol%20Overview-Stefania%20Gazzina.pdf).

The controller Oia Santorinis who narrates her example, in this case, has started working in a busy airport where lots of seasonal (charter or other) flights take place, as it is a high-end tourist destination. Several of those flights are charter flights, low-cost carrier flights, or flights with crew members in the last phase of their on-the-job training or domestic flights that connect different islands. Thus, the culture associated is more service-oriented, for the passengers to arrive and the airliner to depart, than safety-oriented. Air traffic controllers have the local knowledge to be aware of the detailed configuration of the airport, which they were able to cross on foot as part of their familiarization and orientation in the job. Lastly, the small scale creates a community atmosphere for pilots and controllers to develop closeness and mutually appreciate their roles, even when they do not have personal relations.

When discussing the same scenario of Sect. 6.3 regarding the phenomenon of silence, Oia Santorinis relates it with interpersonal communication in the flight deck when silence was attributed to the chain of authority inside the cockpit, as explained in her testimony in the context of multiple factors present and the realities of air traffic control restrictions:

> It is still a fact that crew members, due to their fear of being punished, or humiliated in case of speaking up, do not raise their voice to their superiors when there is a doubtful situation during a flight. Consequently, incidents or, even worse, accidents may occur because of the silence of lower at the hierarchy members of the flight crew.

First Do and Then Repair the Instruction Oia Santorinis is moving one step back from emergency situations and notes that crew members have a fear of being labeled or viewed negatively (humiliation factor) and even get disciplinary punishment when they voice their opinion in front of a superior crew member. It starts simple, in much more mild instances that she labels as *doubtful situations*, a term mostly associated with technical parameters in a flight (e.g., altitude) that might be perceptual errors or interpretations of flight logs. That means that, in such cockpits, silence treatment is cultivated early on and with non-severe actual conditions from those in higher rank or the organizational operation chart. Oia Santorinis goes on to illustrate her argument with an example from her own professional service, quoted below:

> While I was working in Thira island, at the Control Tower, giving clearances to the aircraft take off and land, there was an incident which was caused by a misconception of what I said to the pilot and at the same time by the hesitation of the co-pilot to stand up to the pilot questioning his experience and knowledge. The aircraft, after it was given the route clearance, asked permission to taxi. I gave it taxi clearance with clearance limit the holding point of runway 16 L. I must point out that runway 16 L was seldom used, since the direction of the wind blowing at the area was mainly north. As a result, the pilots and co-pilots of many airlines were not familiar with the procedures of that runway. Therefore, at the case I mentioned before, the aircraft, even though it was cleared to stop and hold its position at the holding point of the runway, eventually entered the runway 16 L.

Thira tower is a service-oriented tower operating with the goal of increasing capacity due to the number of incoming flights that operate on a seasonal basis. Clearances are expected to be arranged in a flowing manner; as a result, misconceptions about the use of the main runway, the waiting spots (as holding points for the next clearance), and the use of auxiliary routes (like the taxiways in which the aircraft moves in a driving mode with its own means) occur periodically. The clearance that Oia Santorinis gave was a conditional one, to "move to a certain point to wait" and not a final clearance to depart. The specific runway 16 L was in use, weather permitting, and traffic demanding and was not expected by the crew as a routine default choice.

Physical configuration of the airport and knowledge of the operating status, in each service period, is one of the advantages of the controller (towards the pilot) as she is fully familiar with the site (even walked through the whole aerodrome, to ensure GA before shift starts). The co-pilot who holds the flight manuals seemed to have received the actual *conditional* clearance when the pilot was expecting the *final* one; but he hesitated to provide his input. Oia Santorinis goes on to explain:

> When that happened, it was summer and high season for Thira, due to the tourist "invasion" arriving by thousands. That day was very busy at the airport and there was very little time for me to watch the aircraft movement in the ground. However, the moment the aircraft entered the runway violating my clearance, I raised my eyes towards it and the first thing I did was to give it the instruction to immediately vacate the runway via the first intersection to the right. Another aircraft was at the final of the runway and had already been cleared to land.

The high season instance and the fact that Thira is one of the world's most popular destinations with visitors that amount 100 times its population get everyone in a rush. Oia Santorinis accepted that she had marginal time to visualize the physical movement of the aircraft till the spot where it exceeds the holding point in the runway 16 L. That was a critical moment as she has given final clearance to another incoming flight to land and the first aircraft had to empty the runway, under her instruction.

It was a *first do and then repair the instructions* instance to avoid a vehicular-like collision (when two planes move as vehicles with their power in the taxiway and collide). How did she explore the situation further?

> After vacating the runway, I told the pilot that I never gave them permission to enter the runway and that I had to file a report because they jeopardized their safety and safety of the landing aircraft. When they came up to the Tower to apologize both the pilot and the co-pilot, and after my shift was finished the pilot told me that he thought he heard me clearing them enter the runway. At that point, the co-pilot admitted that he could have spoken to the pilot when they were wrongly entering the runway, because he was certain that the aircraft was never given permission to get in. According to the co-pilot, he hesitated speaking-up because the policy of the airline was to cover-up in the past similar incidents caused by the same first commander who was thought to be the most experienced of the airline.

Oia Santorinis explains that the pilot-in-command made a severe mistake (not only violate the instruction but also not asking for repetition from the controller and not reading back the conditional clearance). This type of action demanded reporting from the controller's part. Admitting their mistake as a team (pilot and co-pilot) and going to apologize was a decisive step, since they know that miscommunication should be resolved as they will be in the same airport/airplane multiple times. The informal out-of-shift conversation disclosed that the co-pilot had perceived the conditional clearance to the holding spot and not to the runway; but the policy and atmosphere in the cockpit made him hold his voice. Flying with the most experienced pilot in the airline was a professional fence so important that he was willing to discuss with a third party like Oia Santorinis. She was surprised from how talkative the co-pilot was, in her quote:

> The co-pilot was afraid of being humiliated by the first officer and even losing his job in case of a negative commentary of the pilot about his performance [evaluation] during the flight.

It was almost a confessing moment for the co-pilot, as fear of punishment (not getting promoted or having other adverse personal consequences) or fear of being labeled or viewed negatively was like a risk of stigma to him. Another factor that plays a role in this reasoning is the use of retired air force officers in managerial positions, a fact that sometimes enforces hierarchy in nonproductive ways. Simultaneously it shows a confidentiality bonding that exists in aviation community, especially in a small country like Greece, since when something happens it is a "family matter" that need to be discussed, as confirmed by other pilots with whom I have interacted. So, Oia Santorinis makes the main point for why silence was deleterious:

> If I was the co-pilot, I think that I would definitely speak-up because the environment that both pilots and controllers work in is critical and everyone participating should be 100% and certain of what is told, so even lower officers should raise their voice to the pilot and ask the Tower once more to repeat the clearance. As a result, the safety of their passengers, themselves and other aircraft would be undoubtable and never at risk.

Putting herself in the position of the co-pilot, she is definite in her criticism and certain to speak up due to the severity of efficient reception of messages in aviation and the collaborative work needed to protect one another. A middle route to go was to ask for a repetition, so no one would have felt insulted to humiliate the other. In the end, safety goes beyond uncertainty.

Pilot-Controller Communication Loop In this case, intra-crew relations having a top-name pilot with a less experienced co-pilot made the co-pilot to stay shy and silent. Pilots and controllers are equally involved to the ATC system, and any factors contribute to achieve effective radio communication, factors that are not considered in isolation. Flight communication is effective when the receiver interprets the message as an active listener of the transmitted message. Active listening is the process of sending back, as immediate feedback, to the speaker the semantic message that you think the speaker meant both in content and feelings (Gordon 1975). It is a process of the pilot/controller communication loop that connects communication with human factors: perceiving the message, reconstructing the information contained in the message, and linking the information with an objective or an expectation in the coming operating future (like a route, an altitude, or time). That is why, although pilot-controller communication is one-to-one or two-to-one (at a fraction of time), the controller-cockpit feedback is considered a one-to-two or one-to-many communication so that more receivers can sustain the integrity of the communication loop; the other crew members are also receiving these messages.

Oia Santorinis is pointing out a clearance misconception, as there was a limit to a holding point and not an unreserved clearance. Because of the heavy and diverse traffic of that airport confirmation, repetitions are often omitted in the shake of time and radio communication capacity. The use of "runway 16 L" may explain the misconception as an issue of situation awareness (Endsley and Jones 2012; Dominguez et al. 1994). The pilot seems not to have a mental picture of the location and of flight conditions and intentions for aircraft within an area in relation to each other (like incoming traffic and airspace reservations). After the clearance violation, Oia Santorinis used eye communication for a positional situational awareness that asks for the mental picture of the traffic situation in the physical space, with unexpected progressions in mind (the other incoming aircraft). Familiarity with physical space beyond the radar screen prevented any further event but initiated an institutional communication channel for reporting the incident, in a way consistent with a human-centered approach (Leveson et al. 2002).

The ties and dependability developed in the aviation community made it important for the pilot and co-pilot to apologize with their physical presence. The co-pilot admitted that he was hesitant to speak what he heard from the clearance, as he seemed to have been in the same position in the past; he had chosen to contrast the pilot-in-command, but he was humiliated and lost face. The airline policy to protect the *experienced pilot as a corporate asset* was either communicated during the preflight briefings or circulated as practice in grapevine communication (Crampton et al. 1998). The co-pilot's statement seems to introduce an enthymeme in his mind (Katz 1992); it stays resident in his memory, based on his empirical corporate reality regarding the cockpit insights as quoted:

> The pilot-in-command does make mistakes (even if he has the command). Mistakes are undermining the airline's reputation (so individual blame is preferred than organizational one). Airline's reputation depends on the pilot-in-command face (it should be protected by the other crew members, for his shake and the company's).

That syllogism stays in the co-pilot's mind while functioning with goal to avoid potential negative evaluation that will threaten one's career. After all, the co-pilot works in the private competitive sector of the aviation industry, in which employment security is market-driven. Moreover, the co-pilot in the story is willing to disclose those almost esoteric thoughts in the discussion with Oia Santorinis. It is a case of intensity in silence analogous to verbal language intensity (Bradac et al. 1979). The choice to remain silent is not because of neutrality when receiving that incoming message, but because of the perception of the imposition of silence.

Oia Santorinis positions herself in the inquiry by combining a utilitarian and deontological point of view, consistent with the nature of her profession and the content of her training. Here is her quote:

> If I was the co-pilot, I think that I would definitely speak-up because the environment that both pilots and controllers work in is critical and everyone participating should be 100% accurate and definite of what is discussed or told to do.

Being in a supervisory position, since air traffic controllers work for the Civil Aviation Authority[22] of the Greek state, she is definite to voice her say for an accurate message feedback. It is a discourse case of the greater good for the larger number of people and simultaneously the moral obligation to the profession, as safety prevails. And voicing opinion is more a token of teamwork than a challenge to authority; it would just be an inquiry to the tower to repeat the clearance. Feedback confirmation will close the loop.

Silence in the Tower and the Military Setting In the form of controlling the reply to the silence scenario, I have continued the conversation with Oia Santorinis after reversing the question. I asked her whether (or not) she had experienced silence in her own controlling role either as a choice or as an imposed practice. How she proceeds in her response allows a small room for contradiction that is rather justified by the nature of cooperation with military between military and civilian air traffic control (in Thira airport). She begins by setting her context:

> The airport in Thira was used both by the Civil Aviation and the Military Aviation. During military missions, air traffic controllers and fighter pilot supervisors are working side-by-side in the Control Tower. In summer, we used to give priority to the civil aircraft and if we could find some time between the arrivals and departures we were providing service to the military operational flights. The civil air traffic controllers, me included, were not trained to handle military traffic and mainly it was at our own discretion to study the military manuals and "be trained" by the military controllers. Consequently, if there was a mission supervisor at the Control Tower and there was no other civil traffic, they were exclusively in charge for giving instructions to the fighter planes.

The situation in several Greek islands of the Aegean is a bit hybrid, as airports operate quick response alert readiness of stationed fighter planes, as well as host numerous passenger flights for visitors and tourists. Military controllers work in the same control room as the civil aviation controllers and summer schedule prioritize civil aviation flights, with fighter planes in-between commercial departures (in different final altitudes). Training is conducted in different aviation schools, and only informal on-the-job training is realized from a military controller to a civilian one. When the airport participates in national or international drills, there is an additional high-rank pilot in the control tower that serves as mission supervisor. Thus, the whole setting in the tower changes in military-civilian co-working environment. Oia Santorinis goes on to describe her experience in Thira's aerodrome:

> I was one of the controllers who was fascinated by the fighters and had the interest of studying their manuals; however, I was not supposed or authorized at any time to speak to the

[22] Hellenic Civil Aviation Authority was founded in 1926 as an independent branch of Transportation Ministry in Greece. Its mission is the organization, development, and control of the country's air transport infrastructure, as well as the study and laying of proposals to the Minister of Infrastructure, Transport and Networks concerning the overall policy formulation in air transport. Coincidentally, FAA has an institutional history starting on 1926 and grows as an independent branch on 1958 (more data about the chronology is available at https://www.faa.gov/about/history/brief_history/).

> military about their way managing the military traffic. The only case that I could speak up was when there was other civil traffic in order to make sure there was enough separation between the fighters and the civil aircraft. If I believed that the minimum separation between military and civil traffic would be reduced I would coordinate with the military supervisor to prevent any possible mishap. In every other case, I must admit that it was, by all means, a matter of rank and authority that we could not have an opinion or responsibility of whatever happened.

Oia Santorinis is an enthusiastic controller fascinated by aviation and its elite part, the Air Force. The nature of Greek geography and border line with thousands of islands in the Aegean with a dense network of airports and military bases has elevated the HAF forces in the mind of the laypeople and aviation professionals in Greece. In the co-working experience that Oia Santorinis mentions, her major goal was to ensure that the civil aviation routes are not mixed with fighter jets' routes. Safety and the promotion of local economic activities require keeping these flight routes separated and the aircrafts intact.[23] Even smaller incidents (before escalating further) need to be avoided as they affect the sense of security which is part of economic and tourism climate. And it is not an issue of managerial authority but of distinguished professional responsibilities. Coordination was feasible as both parts apply public policies in aviation.

Established conditions of co-working for civil and military flights generate a division of labor and job descriptions that define rank and authority in rigid terms. Separation in flight routes, missions, and service in the air demanded a separation in the air traffic control duties for military and civil flights. Although Oia Santorinis could sound more idealistic in the first part and her enthusiasm for the appeal of fighter aviation is there, she is clear about distinct layers of responsibility. You can speak up for your own (civilian) traffic but cannot be involved in military missions. Consequently, it sounded contradictory in the start, but it is a different assessment of authorization and ranking compared to the pilot-co-pilot hierarchy.

The pilot-controller communication loop mentioned earlier in this section connects communication theory with human factors through a correction-confirmation process. Personal bias or error is introduced in the pilot-controller communication loop. Crew resource management approaches emphasize the relevance of the context and expectation in every communication occurrence. And expectation may have an impact to the effectiveness of communication with a positive or a negative bias; the decision whether to remain silent or not has to do with the concept of missed clearance in the message and the question of authority evolved in a restrictive conflict insight the flight deck. The narrative that was the result of conversation with Oia Santorinis identified the concept of preparedness for the unexpected, for the nonroutine use of the specific airport in an alternative runway.

[23] Vertical and horizontal flight separation standards are laid down by national authorities to facilitate the safe navigation of aircraft in controlled airspace. Observance of these standards ensures safe separation from the ground, from other aircraft and from protected airspace. National separation standards are based on the provisions of ICAO Doc 4444 (Procedures for Air Traffic Management), especially Chap. 5. Differences from these standards (if any) are published in national aeronautical information publications (extract coming from http://www.skybrary.aero/index.php/Separation_Standards).

Also, although the controller is not expected to know the situation beyond what is communicated and what is indicated in the screens, Oia Santorinis went beyond the highly mediated environment in the control tower and activated the direct visual channel that positions the aircraft in the physical space of the runway and not just in a screen (an element of GA). The controller has more resources to administer and has to be able to instruct beyond the daily repetitions, as local situated knowledge is involved for the specific location. An expectation error occurred when the cockpit members thought that they heard the clearance, each one differently. Still their communication has a repair attempt by apologizing, as a gesture of professional courtesy, although a report was filed. Lastly, the co-pilot felt he should not challenge the authority and experience of the pilot, recognizing a corporate practice, and the potential of a trainee-level humiliation for a different opinion is posed.

Since the potential risk of error was more than a probability in this flight case, we need to have in mind the assumption of local rationality; experts even so as practitioners typically do what is reasonable to them according to their understanding of the circumstances (Woods and Cook 1999). Only what was is communicated or not communicated can show their understanding of the circumstances.

5.5 Pilot Silences from Fighter Pilot's Point of View

In this section the same silence scenario is posed to fighter pilots with experiences in attack and fighter planes, as well as helicopters. Since fighter pilots in Greece conduct their duty in the Air Force and Army aviation (the Navy is using reconnaissance planes and helicopters, and firefighting aircrafts operate under the auspice of the Air Force) and they work in pairs and then in larger groups closely together, their responses include specific cultural references for their respective subcultures. Silence occurs inside their cockpits in two-seat aircrafts, or as members of helicopter crews, their communicative behavior resembles the interaction between *the father senior pilot and the son trained pilot*, an association they develop since their early years of training and evolves towards their years of seniority. Several studies have investigated the high desire for achievement, as a motivator for fighter pilots, connected it with the variable of dominance and the military value system with adoptive traits due to the nature of their missions (Triandis and Suh 2002; Ashman and Telfer 1983; Reinhardt 1970; Fry and Reinhardt 1969; Retzlaff and Gibertini 1987).

The sense of mutuality, the value of respect, the exercise of leadership, and the peculiarity of the Army (with norms of conventionalism and abasement) compared to the Air Force (with norms of deference and succorance) are cultural determinants and motivating components of how they conduct their flights with their senior officers in the same aircraft, in dyads or quadruplets. In the sections below, fighter pilots explain mutual silence and mentoring with the senior instructor (Sect. 6.5.1), the leadership and airmanship culture (Sect. 6.5.2), as well as the dialogic communication in checklist marking and the stance towards authoritative passengers and officers.

5.5.1 *Instructor Knows Better, Mutual Silence, Mentoring*

A fighter pilot at the rank of colonel of Hellenic Air Force, namely, *Eleftherios Venizelos*, offers additional insights from the point of view of Air Force flight missions (with A-7 squadrons[24] and NATO Reconnaissance in AWACS[25] aircrafts). He believes that Air Force works because of the groups of people that are calibrated in sophisticated team work after intense training. Eleftherios Venizelos goes back to his early years in fighter planes when pilots are attached to their flight instructors, with a catalytic instructor-trainee relation that is embedded in a deep hierarchical structure.

Instructing Pilot, Authentic Knowledge Eleftherios Venizelos discusses his recollections commenting on the silence scenario. The description is about a common training mission in which the trainee pilot has the first experience of touch-and-go missions several times in the same day in a two-seat training aircraft.[26] It is a supervised flight that emphasizes the preflight briefing and tactical design of the mission taking special care for the fuel configuration and familiarization with as many airports as possible. Souda Bay in Crete is an important destination for these flights; it currently hosts the 115 Combat Wing of HAF with two fighter squadrons of F-16s aircrafts, the US Naval Support Activity in Greece[27] with reconnaissance planes RC-135s among other installations and a civil aviation airport (Chania "Ioannis Daskalogiannis" airport).

He begins by setting the flight context of a training flight that started as a routine mission:

---

[24] A-7E Corsairs were the attack planes used (among others) by the US Navy in the Desert Storm campaign against Iraq on 1991 (details available at http://www.defensemedianetwork.com/stories/gulf-war-20th-desert-storm-was-an-heroic-moment-for-a-7e-corsair-ii/). Then Hellenic Air Force has used them in three combat wings in Greece. When I was serving my duty in 115CW, A-7 s were stationed in three squadrons there.

[25] AWACS acronym stands for airborne warning and control system aircrafts which are modified Boeing 707 s airliners used to locate other aircrafts while being airborne and adjust to reconnaissance conditions. The E-3AWACS component is NATO's first integrated, multinational flying unit, providing rapid deployability, airborne surveillance, command, control, and communication for NATO operations (more at http://www.e3a.nato.int/ and at http://www.boeing.com/boeing/history/boeing/airborne.page). They operate for NATO in Germany, Italy, Greece, and Turkey, and their multinational crew comes from 17 NATO countries.

[26] T-2C Buckeye jet trainer aircraft was produced for the US Navy by North American at Columbus, Ohio. T-2C trainers were used by the Naval Air Training Command to conduct basic jet flight training for future Navy and Marine Corps aviators. The trainer established an outstanding record of safety and reliability while providing training for more than 11,000 students to pilot 18 different models of Navy jet aircraft. Buckeyes also were purchased by Venezuela (T-2D) and Greece (T-2E). Data available from Boeing at http://www.boeing.com/boeing/history/bna/t2.page.

[27] 115 Combat Wing of HAF is described at http://www.haf.gr/en/structure/units/ata/units/115pm.asp and US NSA Souda Bay in Crete at http://www.militaryinstallations.dod.mil/pls/psgprod/f?p=132:CONTENT:0::NO::P4_INST_ID,P4_INST_TYPE:6095,INSTALLATION.

> In a T-2 flight the mission was to fly from Kalamata to Souda Bay with a touchdown and then immediate take-off to return to Kalamata. The plane was equipped with a TACAN[28] beacon-transponder which could project a measurement of forty-degree azimuth[29] error lock-on. Due to the nature of the TACAN signal, it is possible for the TACAN azimuth to lock on in multiple times the 40° from the true bearing with no warning flag appearing. The pilot should crosscheck other navigation aids available to verify TACAN azimuth. The weather was cloudy with low visibility from 16,000 feet till the absolute low altitude. We have executed approach and descent with full TACAN, with no visibility towards the runway or the mountain in the southern of the airport, in IFR[30] conditions, and without cross-referencing the TACAN indicators.

One of the first training flights for the young HAF lieutenants in Greece is the flights from Kalamata where they start having their professional flight practice, before being assigned to fighter squadrons and choose their fighter type. In these missions, a flight to Souda Bay, where the world-leading US Air Force is stationed in Greece, is of an increased mental load for the new-coming pilots as they imagine it as showcasing event. T-2 is a standard aircraft for such training, and the use of TACAN transponder is a navigational tool appropriate for initial pilot or initial pilot orientation; it is an effective position locator when it receives a point of reference from the ground. But low visibility, the mountain, and lack of confidence to instrument indicators led the flight out of the routine. Eleftherios Venizelos goes on to reveal his seemingly reasonable uncertainty:

> While descending there was no communication between myself and my instructing officer, as I wanted to follow the required process religiously, being anxious as to whether the indications were correct or not (without being cross-checked). I wished we would ask for a full TACAN-Ground Control Approach, having the support and instructions of a ground

---

[28] TACAN, which stands for tactical air navigation, is a polar-coordinate-type radio air navigation system that provides an aircrew with distance information, from distance measuring equipment (DME) and bearing (azimuth) information (at http://www.globalsecurity.org/military/library/policy/navy/nrtc/14090_ch2.pdf).

[29] Azimuth in this case is the horizontal direction expressed as the angular distance between the direction of a fixed point (as the observer's heading in the flight route) and the direction of the object (on the airport ground, when other concrete fixed points like mountains are around). Data from http://www.airforce.com/PDOP/. So, azimuth (Az) is the angle measured along the horizon from North. With North being 0 degrees, East would be 90 degrees, South 180 degrees, and West 270 degrees, and finally North could again be described as 360 degrees (extracted from http://www.planetary.org/blogs/guest-blogs/jason-davis/3450.html). When there is radar fixed point on the ground, then erroneous indication is possible.

[30] Instrument flight rules (IFR) and visual flight rules (VFR) are one of two sets of regulations governing civil aviation aircraft operations; Federal Aviation Regulations (FAR) defines IFR as: "Rules and regulations established by the FAA to govern flight under conditions in which flight by outside visual reference is not safe. IFR flight depends upon flying by reference to instruments in the flight deck, and navigation is accomplished by reference to electronic signals." Visual flight rules (VFR) are a set of regulations under which a pilot operates an aircraft in weather conditions generally clear enough to allow the pilot to see where the aircraft is going. It is also referred to as, "a term used by pilots and controllers to indicate the type of flight plan an aircraft is flying," such as an IFR or VFR flight plan (data from. http://shebleaviation.com/study-materials/FAA-H-8083-15A-Instrument-Flying-Handbook.pdf).

> radar. I did not say a word as I thought that my instructor knows better and I focused on implementing the descending process; in my mind I was too young, just out of the Air Force Academy, how could I know it all?

It was a series of touch-and-gos (consecutive lands and takeoffs), and silence was important for Eleftherios Venizelos to keep his concentration and follow the descending steps thinking of a training combat-mission scenario, where communication is muted/sterilized for the period "combat in-combat out" (descending touchdown and then take off and stabilize position). TACAN operates with two reference points and they were missing a clear view from the ground. He started a "what if" inquiry which conflicted in his mind with the relation of instructor-trainee which is close to father-son relation in the Air Force; "*you don't question your 'father' in his instructions*," Eleftherios Venizelos said in our conversation. In HAF's practice senior instructors assume responsibility for and stay with a small group of junior pilots, for certain periods of time; in this sense a strong ("father-son") affiliation is created as if they were taking private classes. He goes on to describe communication interaction with his pilot instructor:

> But when entering to the second phase for the go-around I decided to ask my instructing officer: how do you know that there is no false or erroneous information to the navigation display equipment? With a voice full of relief, he replied that he had exactly the same strain in mind, because of the way we decided to descend but he did not want to admit it to me. I took my lesson and since then and for the past twenty-five years of my Air Force duties I have always been talking through for any doubt I had, when according to my judgment, the safety of the flight mission was at stake. My choice has been, primarily, with the criterion of flight safety and in a secondary manner the potential consequences in my career path.

The fact that they have completed one descend (successfully but with uncertain minds) and they were going to repeat encouraged him to pose the troubling navigation question. TACAN navigation needed point A in the aircraft and point G on the ground. What was different here, for Eleftherios Venizelos, was that the instructor pilot had exactly the same doubt in the conditions they were flying. He was more authoritative by position but, being in the Air Force (unlike what is customary in the Army and less in the Navy), was also able to adjust and be more productive in his training, gaining also the respect needed from the trainee; a mistake admitted and repaired is a lesson learned. Eleftherios Venizelos went on with his flying instinct and made a lifetime decision to speak up when his critical ability drives him to reach safety which is nonnegotiable.

The TACAN radio facility in this usage created a feeling of blindness for the two pilots, as they did not have an instrument to compare flight indicators with actual position locators on the ground. Military discipline and protocol, as well as the strong feeling of *flight instructor that knows all* that is embedded in Air Force culture, prevented the trainee pilot to voice a question. Immediately after, since this mission has several phases with the "touch-and-gos," Eleftherios Venizelos had the eager question that his training forces him to pose; a repeated practice in Air Force missions, learned from the first years of the flight schools, is that you need to ask as many questions as possible during the preflight briefing or even during the flight with the flight instructor.

The first descend was "voiceless" as an application of the sterile cockpit rule explained from "combat entry" to "combat exit," as explained in the introduction

(Chap. 1, Sect.1.3). But the inclination to resolve the doubt triggers the survival instinct that pilots develop via real-flight mission training[31] out of simulators. The reply was enlightening as the instructor pilot opened up to confirm the doubts, as his own. Eleftherios Venizelos has identified a case of mutual silence in communicative behavior in the same scenario. It relates to the cognitive bias of an implicit personality theory, in which a trainee co-pilot's followship is attached to the instructor pilot who has the leadership and the experience which make him untouchable, especially since we speak about military aviation that starts with a boot camp mentality (Eden 1992).

In Eleftherios Venizelos's narrative, it is evident that the flight had a detailed briefing preparation from the trainee pilot, whereas the instructing officer may have considered it a routine flight, in a familiar destination. Military chain of command was a restricting factor for their communication, especially because the trainee pilot was just out of the Air Force Academy with the mindset of connecting the commanding officer with authentic knowledge. The Air Force culture, especially in flying personnel, is adjustable to a more flexible code of not "who says" but "what you say." That is a bypassing route that enforces trust in the chain of command, sometimes more efficiently than mere ranking.

I had the same experience as an airman and then a petty officer-programmer in my service both in the Greek Pentagon (with brigadier generals and higher ranks) and in Air Force bases (with active fighter pilots). In a retrospective evaluation, I could agree with the numerous discussions I had with different commanding officers of mine when I asked them to comment on what is different in the Air Force. They claimed two reasons; pilots value each single day, because it might be their last and the sophisticated bond of pilot-aircraft union helps them understand human complexity and the complexity of the real world.

On the other hand, the instructing pilot had the primacy effect of *just another training flight to Souda* with the self-serving bias of controlling the situation. Even when doubt came in his mind, the attribution of control and hierarchy did not allow him to communicate symmetrically with the trainee. The resolution came when the question posed confirmed the realistic nature of the situation. As Eleftherios Venizelos explained in my probing question, the instructor's acceptance of the doubt and risk and his recognition of a flaw in the flight design during the debriefing phase of the mission (upon their return) reinforced the instructor-trainee relation into a mentoring relationship in the years to follow (Ragins and Kram 2007). Furthermore, the situation was a very good reminder of how relevant their safety training was and that established a paradigm in his future behavior; safety is first, for the personnel, the plane, and the mission.

Military Controller's Point of View Eleftherios Venizelos reply was about intra-cockpit communication in a two-seat fighter plane. As a control filter for his reply, I posed the question of silence from the point of view of military controllers and whether they tend/want to speak up or not. Eleftherios Venizelos pointed out another hybrid case differentiating from previous responders. In HAF missions there are an

[31] Expressing the thoughts for that question resembles the use of verbal language as triggering instincts to solve "evolutionary significant problems," in Pinker's approach of the language instinct.

air traffic controller and a tower supervisor who is a higher-rank experienced pilot. The controller is dealing with the basic air traffic control, and the supervisor is there just in case an incident occurs to facilitate approaches, takeoffs, and landings but with no authorization to intervene except for the cases that he is asked for mission assistance from the fighter planes that are airborne.

The tower supervisor is not a substituting pilot, and still responsibility lies on the actual flying pilot who needs to take the final decision, as also emphasized for the pilot as decision-maker by Kochan (2009). Despite these strict divisions of roles, Eleftherios Venizelos recalls interventions in traffic management issues from higher-ranking officers that were functioning as supervisors, using the military hierarchical power and not the actual position description.[32] This supervisory position rotates in each squadron and contributes to an increased mission awareness and effective briefings: the today's supervisor is the tomorrow's mission chief and vice versa.

5.5.2 *Safety Culture, Trust, Situation Awareness, and Leadership*

The last set of responses discussed in this chapter covers the responses of two other fighter pilots who a more institutionalized point of view of silence due to their extensive experience in international missions (for the first case) and fighting helicopters (for the second case). Posing the same scenario regarding silence, the two pilots were able to recall latent voice episodes (Bienefeld and Grote 2012), i.e., ambiguous situations in which crew members had felt that speaking up was necessary for flight safety and how they have seen that evolving during their career. As they have been serving as Chief Training Officers and Squadron Commanders, they have been in both upward and downward voiced communication directed to lower or higher level of hierarchy, respectively.

In this fourth discussion, conversation takes place with *Ioannis Daskalogiannis*, a HAF fighter pilot at the rank of Brigadier General with 20 years of operational experience on board of T-2s, A-7s, and Mirage 2000–5,[33] which are considered the elite unit of HAF. Ioannis Daskalogiannis' approach on silence is stated below. What is different here? The conceptual understanding of human behavior in a certain organization setting which encourages collective orientation:

---

[32] Similar events were projected in the "Silence in the Air" (Anosis SA, 2000) series of the Greek national TV channel "Mega TV" that were mentioned in footnote #14. The stereotypical scenes were describing interventions in the work of female controllers or engineers. The series was another example of the strong affinity that Greek society feels with their aviators that defend the Greek FIR and how their Air Force routine governs everything else in their life.

[33] The French Dassault Aviation explains the technology and equipment of Mirage 2000–5 at http://www.dassault-aviation.com/en/defense/customer-support/mirage-2000/mirage-2000-5-mk2/.

> In my opinion this is not an issue for the crews and not an issue driven by the particular human behavior. This is an issue solely driven by the safety culture level of the organization itself. At low levels, (i.e. pathological and reactive) the lack of trust among professionals, on the possible consequences of their actions, encourages the 'not speak up' behavior. At higher levels (i.e. calculative, proactive and generative) of safety culture the increasing trust encourages communication and a 'good' CRM behavior. As a fighter pilot for more than 25 years, I found myself in all different organizational safety culture levels and therefore I also found myself in all different behaviors (i.e. hiding, fear, not speaking up, speak up).

Safety Culture and Silence Ioannis Daskalogiannis speaks about organizational safety as a cultivation process which advances with the experience of events and the training patterns, from a low level that is apathetic to a higher level which is constructive. His personal path went through similar evolution while he was adding flying experience and theoretical foundation in different schools. The perceived behavior generates a ladder of confidence over time (and with service) which is compatible with trust and covers the states from hiding till the tendency to speak up. What does he suggest in order to accomplish that?

> One solution could be to lead by example and this lead approach should be a top-bottom approach. The experienced aviation professionals should be the first ones that report their slips, lapses or mistakes experienced during their execution of their duties and then they should ask from younger professional to follow up with their own perceptions about their human performance. This will promote trust, airmanship, higher safety culture levels and a speak-up policy, over time.

This is his second contribution to suggest an effective strategy to break silence barriers with the confidence to speak up, leadership by example. The leading aviator knows how to learn from his mistakes and the mistakes of others by reporting his own first. Ioannis Daskalogiannis makes two important points: (i) institutional commitment to encourage speaking for safety and (ii) the quality of leadership which may expedite the movement to the ladder of speaking with a trusting atmosphere.

Ioannis Daskalogiannis emphasizes the collective orientation of group work that cultivates behavior beyond the individual, by taking an educational, managerial, and organizational point of view. He takes the incidents of silence as symptoms that derive from deeper causes, not just the particularities of attitudes of different individuals in the aviation vocation. Safety culture is the goal but not a static one, as it develops and evolves with time, with training, when incidents or accidents occur, and with interpersonal and organizational trust. Two are the key factors in his analysis to prevent silence from happening: safety culture as explained and imposed by effective crew resource management, as a form of organizational communication, and leadership in taking the example from top to bottom.

Fighter pilot's culture in HAF was formulated in the past 30 years with the daily real-life training in interceptions over the Aegean Sea, NATO tactical schools and missions, the relevant accomplishments in drills and international flight schools, the paradigm of self-improvement in international standards, and the application of meritocracy in their selection process. Apart from that, their presence in the Greek periphery outside the main cities has sculpted ties with the everyday Greeks who approach fighter pilots as defenders which provide in national security and territorial integrity, including the Greek Exclusive Economic Zones[34] (Jacovides 2012; Oral 2009; Kariotis 1997). These are perceived ingredients of a microculture as a subset of the broader military culture which has historical roots since Greek antiquity. This aspect of culture communicates respect and pride to the insiders and the outsiders. It is a concept necessary to explain group and organizational communication in Ioannis Daskalogiannis' terms.

Pilot Microculture and Perceptions Culture is crucial in communication theory and in building a knowledge infrastructure, as it is infused in contemporary analyses on all academic levels, including the Air Force Academy (Jandt 2004). This type of bonding coupled with social factors has created an elite culture, in a positive manner of professionalism, higher skills, and commitment to duty with acceptance of a life risk as if it was an ordinary attribute. HAF's fighting personnel (and their mechanics) has gained this status after several pilot casualties/sacrifices on duty and captivated the admiration of the public (Tsolakis 2013). Despite the elite character in HAF's culture, when one speaks with pilots, they are inclusive and accommodating and don't account and do not project themselves for a superior culture relative to other subcultures (Berry et al. 1992). The modesty that pilots manifest in their routine life and contacts with the lay person amplifies the public image, welcoming impression, and pride of their service (as they are perceived as professionals with patriotic duty and competitive tasks).

They are aware of their status and of their vocational differences, but they value their everyday interaction with others and retain homophily ties among their colleagues from the same year in Air Force Academy and the same training squadron. Being in Air Force Academy together, assigned to the same squadron together, service in the same combat wing together, standing by in readiness together, short-term stationed in another base together, fly in dyads or quadruples together, or fly in a two-seat aircraft are landmark events in the life of Air Force pilots and in HAF.

Secondary, but collaborative, experiences are the flights with ground engineers (who oversee the aircraft) together and with controllers (who oversee navigation with them). In all these stages, the use of pronoun "we" might sound unstructured during the years in HAF Academy but is enacted with the sense of "we-ness" (Handelman 1990; Sturken 1998; Turner 1969) and a conception of community (Bruhn 2005).

---

[34] The United Nations Convention on the Law of the Sea (1982) has crafted the legislation for zones of economic activity which extends beyond the intangible borderline in the sea (surface, seabed, and continental shelf in a middle line between adjacent states). It is how economy "reads" the maps of border protection, rescue zones, and exploitable resources in territorial waters (as explained at http://www.un.org/depts/los/convention_agreements/texts/unclos/part5.htm.). Greece, for another time, is a distinct case due to the island character of its geography(see Ministry of Foreign Affairs positions at http://www.mfa.gr/en/issues-of-greek-turkish-relations/relevant-documents/delimitation-of-the-continental-shelf.html), whereas Turkey is arguing on the closed Aegean Sea which delimitates different borders in mainland and in the islands (rather arbitrarily) as explained from Turkish Ministry of Foreign Affairs at http://www.mfa.gov.tr/the-breadth-of-territorial-waters.en.mfa. The question of international law for the maritime environment is analyzed in the respective references.

Both pilots and controllers cope with flights and crisis together, in the phases mentioned above. It is a lifetime bonding, as fighter pilots-responders, confirmed in our discussions. When pilots change "mates" or partners in one of the above phases, it takes them some time to adjust in the new subculture, although adjustment is part of their predefined career path that type of bonding is recreated.

Culture consists of relatively specialized practices and elements of a lifestyle or professional lifeworld that are passed through communication (and not biologically) from one generation to another. Hofstede (1980, 2009; Hofstede and Hofstede 2005) spoke about a collective programming of mind for thinking, feeling, and acting; Hofstede proceeds to analyze culture at an organizational level when he introduced the dimensions of culture[35] that are evident in group work where subcultures exist: power distance (to explain perceptions of authority), individual and collective orientation (towards the teamwork needed for any flight mission), masculinity (that may explain gender differences in aviation industry), uncertainty avoidance (that may explain lack of decision-making), long-term orientation (that may explain actions without situation awareness), and indulgence or restraint (that may explain the human drives of airmanship and the sense of freedom in the air).

Based on its institutional configuration, HAF as an Air Force is a knowledge-centric organization which applies managerial patterns, interpersonal ties, and knowledge exchanges of the most effective organizations.[36] Taking the aviation professionals as an organizational category (Merrin 1996), several subcultures exist in a form like communities of practice for air traffic controllers, pilots, and crew members, and microcultures develop for fighter pilots and personnel of military aviation. In Wenger's et al. (2002) analysis, communities of practice are groups of people like the above who share an interest or a passion (aviation is a passion for aviation professionals, as almost all responders stated) for something they do (their part in flights) and learn how to do it better (understand the role and challenges of the other in the flight) as they interact regularly (even outside the conventional work schedule). Lesser and Storck (2001) describe a community of practice as a social entity with, informally bounded members, who share and learn, based on common interests through their mutual engagement. Furthermore, this entity develops a social learning engine which enables the members to collaborate, share ideas, find solutions, enrich their tacit knowledge, and create adaptive solutions, like heuristics. Previous studies on managerial aspects of communities of practice support them as

---

[35]At an organizational level Hofstede's dimensions are broken down to means-oriented vs. goal-oriented, internally driven vs. externally driven, easygoing work discipline vs. strict work discipline, local vs. professional, open system vs. closed system, employee-oriented vs. work-oriented, degree of acceptance of leadership style, degree of identification with your organization. Hofstedes's theory and applications are further explained at the senior's site at http://geert-hofstede.com/organisational-culture-dimensions.html.

[36]A typical example of the immediacy and efficiency of HAF (despite the economic crisis period) was the accomplished coordination and precision accomplished in the recent national drill "Iniohos 2014," functioning with the "hot seat" principle for a fighter jet to land, refuel, reequip, and take off again during the 4 days of the most competitive aerial drill in Europe. Another 800 fighter jet takeoffs (day and night) were reported for HAF (at http://www.haf.gr/el/news.asp?id=8703).

a dynamic communication resource, which operates beyond typical frontiers, based on internal mechanisms of communication but also based on leadership and trust (Loyarte and Rivera 2007; Roberts 2006; Chalmers and Keown 2006). The shared practice of constant improvement and learning the flight generates the norms in the subcultures of this community.

These microcultures are bounded by the rules, routines, behaviors, and attitudes, as cultures within cultures (Hayward 1997). Safety culture that Ioannis Daskalogiannis states is one of those *cultures within the culture* and is inextricably linked with the organizational culture at an occupational or operational level. Safety culture[37] (in this case microculture for fighter pilots or controllers) is the way safety is perceived, valued, and prioritized in an organization. Organizational, occupational, and (very often) national cultures are positive, negative, or neutral determinants of a certain safety culture. Its collective orientation is underlined with the inquiry *how does this organization/group behave when no one is watching*? Safety regulations are implemented successfully when a "just culture" reporting environment within aviation organizations, regulators, and investigation authorities is attained. Operating in just culture conditions means that individuals do not take a blame for "honest errors," but are held accountable for deliberate violations and punished for gross negligence. The resultant effective reporting culture depends on how those organizations handle blame and punishment and the gravity they attribute to human error (*a mistake*), at risk behavior (*a chance of harm to occur*), and reckless behavior (*an action that knowingly puts themselves or others in harm*).

As safety culture aims to locate erroneous situations wherever they happen, it is communicated via CRM, starting from the cockpit to the crew, as explained in the start of this chapter. CRM is a process of enculturation, using all available resources for flight crew personnel to assure efficient operation, error reduction, and stress avoidance. CRM was developed in light of the new insights from aircraft accidents after the introduction of flight data recorders[38] and cockpit voice recorders[39] into modern jet aircrafts; information gathered has suggested that technical malfunctions, failures of aircraft handling skills, or a lack of technical knowledge was only part of the causes. But, the inability of crews to respond appropriately to the situation in which they find themselves was another important parameter. That means that human fallibility is recognized and routine errors are not punished but used for a systemic review of the flight operation in the organization (Leveson 2012; Loukopoulos et al. 2009). In such an environment the organizational response via

[37] In the category of organization and human performance, Skybrary repository analyzes the components of safety culture and the connected just culture (at http://www.skybrary.aero/index.php/Safety_Culture and http://www.skybrary.aero/index.php/Just_Culture, respectively).

[38] Flight data recorder (FDR) – device used to record specific aircraft performance parameters. The purpose of an FDR system is to collect and record data from a variety of airplane sensors onto a medium designed to survive an accident (ICAO Annex 6 "Operation of Aircraft", Vol. I).

[39] Cockpit voice recorder (CVR) – a device used to record the audio environment in the flight deck for accidents and incident investigation purposes. The CVR records and stores the audio signals of the microphones and earphones of the pilots' headsets and of an area microphone installed in the cockpit (ICAO Annex 6 "Operation of Aircraft," Vol I).

CPM is to reduce the likelihood of error, to trap error before it affects operations, and to mitigate its consequences when error really occurs.

Leadership, Trust, and Airmanship We now finally continue to the social point about leadership. Ioannis Daskalogiannis' prescription is to cultivate responsive leadership, especially because he speaks from the inside of the Air Force. In organizational communication it is accepted that you cannot lead with your feet on the desk, but you need to stay involved (Fuller 2011). Leadership emerges as the process of influencing behaviors and establishing directions to be followed, on the one hand, and empowering others, on the other hand. The leader should listen effectively and encourage the others to speak and overcome their silence (Johnson and Bechler 1998). Knowledge, approachability, and dependability are substantial traits of leadership and start from reporting your own slips or mistakes (Hackman and Johnson 1991).

Ioannis Daskalogiannis states that trust (like respect) relies on mutual give-and-take (Fuller 2011). The leader has the ultimate responsibility for forming a strong bond of trust with the younger professionals to inspire them to "*follow up their own perception about their human performance*," as Ioannis Daskalogiannis emphasizes from his vast experience in active (not desk) leadership roles in HAF. Controlling the natural fear of flying and building confidence is a self-assessment process for younger fighter pilots (and not only). They evaluate their own performance through the constant mission training in dyads, in two-seat fighters, and in quadruplets and in-flight simulators/emulators when they change their flight status. In most cases this self-assessment process is conducted during the readiness phases (for day or night missions) and while on board the cockpit for a quick response alert for 5 min; both cases are usually repeated every day and every week and fighter pilots are thinking aloud with themselves. Each mission has rigorous briefing and debriefing and posts a major objective and alternative goals (to pursuit when the main objective is unreachable), during the tactical design and preparation.

In further discussion, when the question is reversed about his personal experience to speak up in his own roles, Ioannis Daskalogiannis is quite consistent in his narrative:

> I preferred silence only when I did not have Situational Awareness (SA). The rule of thumb is: The one who has the SA is the one he finally speaks and makes the decision. The others follow the command. In general terms the one that has SA has the authority whatever the rank.

The flying conditions in HAF missions where being airborne is the typical case, unlike other Air Forces which fly mostly in international missions, are intense, and multiple inputs are present in every mission. The coordination of these multiple stimuli determines the perceived definition of the situation while flying, in a dynamic way. The one who acquires the situation and connects it with the tactical design and the coming actions of the near future during the mission is the one who speaks, if even he is not the first in rank at any specific moment. In the next moment the message is relayed and observed using the mission command and the broader chain of command.

In any flight formation, cohesion is sustained in order to fulfill the mission goals successfully. When probing for this question, I have asked Ioannis Daskalogiannis to recall incidents from his long experience in interceptions and dogfights against Turkish fighter planes flying over the Aegean Sea.[40] It is quite striking that he confirms that SA was the key issue even when quadruplets or double quadruplets were in defensive missions. These formations had to follow their readiness briefing and the burst of communication with intra-formation signals and real-time transmissions from National HAF Operation Center: the pilot who has the best grasp of the current situation is speaking up and the others follow.

The next step is to apply the rules of engagement which are a responsibility of the mission chief, as is a matter of judgment, decision-making, and compliance. For Ioannis Daskalogiannis, "SA-first" sentence was a feed-forward in his briefings and a closing statement in his debriefing and an ethical responsibility to be a responsive leader that empowers others, even in such an operational environment where rank must comply with operational competence (Bennis and Nanus 2003). This is the airmanship culture, as Ioannis Daskalogiannis puts it.

5.5.3 Conducting Checklists, Morals, VIPs, Arbitrary Non-Reporting

The closing report concerns a pilot of Hellenic Army Aviation at a rank of a Colonel, namely, *Ippokratis Koos*. He is specialized in Chinook[41] helicopters and serves with close cooperation with Apache attack helicopters.[42] He experienced a chain of command established without flexibility, a tradition of hierarchical organization, and ethos that goes back to ancient historical warfare of the epic works of Homer, Thucydides, Xenophon, and others, as described in Pressfield (2011) and Rich (1995), as well as Allison (2017).

When transferred to helicopters, he found it revitalizing being always on the run, with multiple readiness alert missions, both for the military and civilian purposes like distinguished guest transport or search and rescue flights. In the most

---

[40] The Hellenic National Defense General Staff is monitoring all violations of National Airspace and Infringements of Air Traffic Regulations (ICAO) inside Athens FIR. All violating aircrafts are identified and intercepted according to the relevant rules and procedures of international law. The listing is publicly available at http://www.geetha.mil.gr/index.asp?a_id=2779. A detailed explanation of one of these cases that escalated to a crisis is analyzed in Katerinakis, T. Drexel Graduate Portfolio (2009), regarding the Imia crisis available in Drexel Hagerty Library at https://idea.library.drexel.edu/handle/1860/3297. In 2013, during the national military drill "Parmenion," HAF has implemented 800 takeoffs in a day over the Aegean Sea.

[41] The Chinook is a multi-mission, heavy-lift transport helicopter further explained from Boeing at http://www.boeing.com/boeing/rotorcraft/military/ch47d/.

[42] The multi-mission AH-64D Apache Longbow helicopter is the next-generation version of original AH-64A Apache further explained from Boeing at http://www.boeing.com/boeing/rotorcraft/military/ah64d/.

common current operational formation, Chinooks fly with three (pilot and co-pilot and one flight engineer) or five crew members (pilot and co-pilot, two flight engineers and a crew chief). Also, they need to adhere to generic air traffic control and to the command control of the Army Aviation that is separate from the Air Force's air traffic control.

When the silence scenario is set for discussion, Ippokratis Koos had four different real stories to share:

> There were times, as young UH-1H helicopter co-pilot, when I was trying to read the checklist steps and the "fully confident" pilot in command, was completing all procedures by memory. Sometimes, I tried to remind him of something he has failed to check, he answered with arrogance ("I already checked it"-though it had not been checked). Such situations, lead me into silence or at best to reluctance.

Checklists and CRM What is the point he is making for teamwork here? Ippokratis Koos as a young second pilot was following the procedural steps by the book, in order to facilitate the pilot-in-command and have a point of reference for the standard procedures and for the airworthiness of the helicopter. If a checklist entry is not really operating (and for that reason checked), then the crew needs to evaluate the condition with predetermined criteria about what is not working. The pilot-in-command had already constructed a routine in his mind to check the listed indicators and controls by heart.

As Ippokratis Koos was a member of the same crew in multiple missions (that was the quantifier "sometimes"), he noticed the deviation, items that the pilot missed from his memory, so check marks that were not representing the observed condition. Then as he wanted to avoid an incomplete checklist and the phenomenon of false memory for the pilot-in-command, he pointed out the missed entries. The pilot did not want to admit it and adjust in the next mission but posed the arrogant remark "I have checked it" knowing that it was not truthful as Ippokratis Koos was monitoring the list. "If I have the right with me and speak up, I expect at least a reservation from the pilot, but not a definite negation of reality. So, do I insist?" Ippokratis Koos was thinking aloud when I moved the conversation on. That was a motivation pushback for the young officer who wanted to cooperate effectively in that crew.

Following the checklist is a default standard operating procedure that is expected in any flight; it is the foundation of pilot standardization and cockpit safety (Degani and Wiener 1993). Also, checklists are a major defense mechanism against equipment failure and pilot error but have cognitive limitations when checklist responses become automatic and not actually crosschecked as verbal communication instructions (Dismukes et al. 2007).

The *challenge-verification-response* mode when the pilot flying configures the aircraft from memory and the co-pilot uses the checklist to verify the accomplished entries breaks automaticity, and checking becomes more deliberate (Dismukes and Berman 2010). Chinook helicopters have a long-serving period of reliability but also a rich history of incidents worldwide, and their checklists are well adjusted to project realistic conditions. A *flow-then check process* asks for scanning the cockpit panel in systematic flow, setting, and checking each item in turn and then the associated checklist. The flying pilot who was the pilot-in-command devolved the checklist to "check only." This fact increases the importance of not just marking the checklist. On the other hand, flight realities report incidents where confidence is overestimated by experienced pilots who take it in the form of explicit pieces knowledge that if missed can be repaired by the available manuals (Doak and Assimakopoulos 2010; Loukopoulos et al. 2003).

The so-called arrogant behavior from the experienced pilot may be a decision to follow tacit knowledge in his memory instead the one verbally offered. Reading the checklist and following the checked notes implements an intra-communication mechanism that verifies the situation. It is a sort of a dialogic interaction with repetition and confirmation which facilitates the checklist monitoring; one says or asks the other performs the action and the former checks the mark. The motor movements of the use of hands to touch or point to appropriate controls, switches, and displays (from the flying pilot) while conducting the checklist (from the non-flying pilot) is a good practice to avoid interruption gaps in fulfilling the checklists.

From a CRM point the focus is on coordination, social interaction, and combined performance in the flight cockpit crew (Foushee and Helmreich 1988). Moreover, co-pilot's reluctance to challenge the flying pilot, when conducting checklists, is consistent with flight simulator studies executed to prevent undesired aircraft state (Orasanu et al. 1998). Also, in most cases, the checklist philosophy of use is the outgrowth of the airliner culture in civil aviation and the camaraderie in military aviation; the Army aviation is less tolerant to assertiveness from a junior officer when compared to the Air Force (like in Sects. 6.5.1 and 6.5.2 for Eleftherios Venizelos and Ioannis Daskalogiannis, respectively).

The set-up redundancy embedded in the dialogic format of a checklist monitoring streamlines efficiency and supports crew collaboration. The two-layer redundancy includes (i) the configuration of the aircraft from memory and then verification in the checklist and (ii) pilot- co-pilot monitoring while conducting the checklist procedure (mutual redundancy). Army aviation is not so accustomed in the use of monitoring checklist as a communication tool (but more as a "to do list") due to the persistent hierarchy, the nature of their equipment, and self-sustained warfare. The status difference and the unwillingness to accept what was missing create a safety flaw and discourage the young new pilot, as Ippokratis Koos was. When I followed-up in his narrative, he noted that in the following round of personnel evaluation that pilot was transferred in back-office work covering an administrative position (not only because of such events).

Hierarchy and Self-Censoring In another instance, Ippokratis Koos describes a self-censored occurrence which was self-imposed. The pilot-in-command thinks one way as superior even when the map shows otherwise:

> I demonstrated similar reluctance to mention a derogation of the preplanned flight path (tactical flight below 200 feet), as the pilot in command was one of the most experienced in the unit and in accordance with the "morals" of the time he was not seeking the help of his copilot. The result was a wrong track for some miles until I find the "courage" to say the derogation in accordance with what I was reading on the map.

What happened in this event? Ippokratis Koos was holding himself not to speak up as he was flying with one of the most experienced pilots in his formation who has the ethos of the speaker, as the higher-ranked officer. The informal expected practice was to wait until asked, as a moral obligation in the microculture of this unit. As the error was derogation from the flight plan, it could affect the operational capacity of the crew members to complete the mission. Ippokratis Koos decided to speak up when the diverted route became evident and the factual reality on the map provided no room for other interpretation. The map was the encouraging tool for a confident assertion to speak up.

Army aviation is organized in more inelastic and conventional personnel relations, in which rank has the authenticity of the insignia and no dialectic inquiry is accepted. This fact creates an informal code of practice for accepted behavior; another microculture enacted inside the helicopter cockpit. When the derogation was realized as happening, it was the opportunity for Ippokratis Koos to voice his opinion from the map. Again, rank superseded reality. In my conversation with Ippokratis Koos, it seems that he was not attributing the no-talk to the superior officer but to the censorship of self, as it happens in instances of negotiations (Schafer and Crichlow 2002). The microculture of groupthink of their unit (the Army Aviation battalion where the helicopter functions) was perceived that way, because of in-group pressure (Janis 1983). Moreover, it seems that this censorship was creating mission uncertainty even to the pilot-in-command as he was not receiving the collaboration he was expecting from the lower-rank co-pilot that was Ippokratis Koos.

Ippokratis Koos was certain as he told me that they discussed the whole incident during the mission debriefing and the experienced pilot prompted him to speak up earlier rather than later. The pilot-in-command was not asking and Ippokratis Koos was not telling, until after the fact.

Evolving as Pilot-in-Command In another case Ippokratis Koos is moving up as a pilot-in-command and is in the position to accommodate everyone to speak up, by having the rank to determine communication flow. But what happens in the VIP mission?

> Later, as a more experienced pilot in command of CH-47D, under the pressure of a VIP transfer mission, I did not mention to the landing area ground-staff the inadequacy of the field due to dust from the downwash, because of the field was not been prepared wet, as stipulated in the regulations. The result during landing was a limited brown-out conditions experience.

Moving upward in the rank, Ippokratis Koos became the experienced pilot-in-command who seemingly had no reservation from others in the crew. But he had a very important passenger in his VIP transfer mission, in a flight out of his military context. It was a head of state from another country in an official visit in Greece who wanted to visit the territory of Mount Athos close to Thessaloniki area, as Ippokratis Koos remembers in our discussion. The extraordinary focus in this flight transformed it to a mission of service and not of operations. The VIP was accompanied by the Chief of Hellenic Army who gave the order for this transportation flight.

So, Ippokratis Koos was flying an extraordinary transport mission for which he was paying attention to the service aspect. The landing area in such flights needs a washup to facilitate visibility from the dust and create a comfort feeling for the passengers. In helicopter flights, depending on the number and configuration of helixes and the configuration of the landing area, the landing spot is chosen or adjusted to minimize the risk of foreign objects damage from foreign objects debris[43] which is quite common in such cases and is regulated in flight manuals. In this flight, Ippokratis Koos found out that the landing field was inadequate not so much for the helicopter but more for the sense of comfort for the passengers. Still they experienced reduced flight visibility due to airborne particles without any other consequences at the time, except for a brief comment from the Army Chief who has noticed the light brownout.[44]

The pressure for time and service due to the "civilian" nature of the flight prohibited Ippokratis Koos to speak up although he was the chief pilot; and as he pointed out, in our discussion, he would expect another crew member to provide a reminder, but they did not have time for a thorough briefing. Social pressure and the change in the mission standards because of the distinguished visitor made him to remain silent even when he had the full authority and responsibility for the flight. It is a perceived conflict between efficiency/passenger comfort and safety event since a shift from military operations to VIP transfer disturbed his attention (Bienefeld and Grote 2012).

In his concluding example, Ippokratis Koos received silence from his crew in a potential mechanical failure. There is a division of labor among crew members, with special care for in-flight engineering and weaponry use. But each crew member has the obligation to check and report to the pilot flying the helicopter the situation in his area of control. So, what is he adding about silence?

> The worst happened when a flight engineer did not report to me, until the end of the flight, a "chip detector" indication on his ramp control panel about a gearbox, because "in his point of view", it was a false indication and he did not want to disrupt the mission. In any case, in accordance with the operator's manual, the pilot in command should be updated instantly on the indication and perform the proper procedure.

As helicopter pilots may fly in a short (often) unscheduled notice for multiple missions beyond their original training, Ippokratis Koos has experienced several manifestations of silence, with diverse or mild consequences. This is the worst case as

[43] All major manufacturers issue special reports regarding foreign object debris (FOD) at airports or landing spaces. FOD includes any object found in an inappropriate location that – because of being in that location – can damage equipment or injure airplane or helicopter and their personnel. Usual recommendations are sweeping and washing and use of magnetic bars, rumble strips, and FOD containers. This is an example from the Boeing company at http://www.boeing.com/commercial/aeromagazine/aero_01/textonly/s01txt.html.

[44] NASA's helicopter brownout research describes helicopter brownout as a dangerous phenomenon when helicopters land in dusty, wet, or toxic environment "whereby sand or dust particles become swept up in the rotor outwash and obscure the pilot's vision of the terrain. This is particularly dangerous because the pilot needs those visual cues from their surroundings in order to make a safe landing" (data from http://halfdome.arc.nasa.gov/Research/Programs/brownout.html).

silence had a bad timing and lack of reporting during a flight mission. The flight engineer located an indicator flashing detecting a potential electrical circuitry problem. Instead of reporting it to the pilot he kept to himself to examine it technically but without reporting it arbitrarily.

The most striking for Ippokratis Koos, as the pilot-in-command, was the flight engineer who assumed the responsibility to decide (with his own point of view) that it was a "false indication"; the flight engineer was withholding substantial information. The flight manual demands intra-cabin real-time communication and more importantly for mechanical issues which may jeopardize the airworthiness of a trusted piece of equipment. It is clear that the engineer realized his mistake (while doing it), since he was anxious to report his decision in the end of the flight during the debriefing phase, as Ippokratis Koos noted in our further discussion.

It was an erroneous behavior and action (not to report and to decide on the severity without reporting) synonymous to poor individual performance (of the flight engineer) which could have an impact to the collective operational goals of the mission. Unlike other errors, which are mostly a post hoc social judgment which cannot be defined in advance of happening, the flight engineer distorted the decision-making process of the pilot-in-command as Ippokratis Koos was governing the helicopter based on insufficient information on their situation. In several cases, since military officers cooperate in multiple obligations and missions in pairs or small groups and roles change, there is a potentiality of gunnysacking (Bach and Wyden 1968); past grievances from one mission may conflict with current grievances for another. Silence in these cases is used to avoid conflict but undermines safety.

Safety is not inherent in any system, neither flights. People must create the conditions for safety in the way they operate a system with multiple purposes; the flight engineer disrupted the dependability bond which holds the crew together (regardless the chain of command). Ippokratis Koos commented to me, "how could I fly safely when I know that my crew may know something that they are not telling me?"

In this case, Ippokratis Koos did not restrict the crew members to speak up; the flight engineer showed lack of situation awareness at a macro-level (as LA and TA, after understanding the conditions, but not projecting the future actions appropriately) by interpreting the elements in the helicopter within a volume of time and space of the mission, applying individual judgment beyond the comprehension of their meaning and the projection of their status in the near future. It was a violation of SOPs which was reported as silence and could be distracting in this case, since that is why standard operating procedures are there for. Ippokratis Koos was definite that speaking up should be based on observed reality and SA and not on rank and authority. Ippokratis Koos was more concerned about the trusting gap in organizational culture which he was trying to cultivate at a battalion level and at the micro-level of his crew.

His decision was to make it clear on his briefing that he valued the opinion and assertiveness of the crew members; but, the crew has the operational duty to convey the information about what they monitor to the pilot flying who is in command and has the responsibility to take the final decision. Ippokratis Koos did not want his crew to feel an unnecessary institutional gatekeeping barrier (rank should not be an open "gate") which would interfere with their operations, in the original sense described by Lewin (1947).

5.6 Knowledge-Worthy Silence Occurrences

This chapter presents the multidimensional concept of silence as understood and used from several aviation professionals who discussed their experiences. Experiences regarding silence usually mark their perception or the context of their work and are determinants of their decision-making. This chapter offers a structured approach to look at silence in an applied domain. Different interlocutors, in this study, perceive silence differently in their roles and connect the phenomenon with diverse practices, regulations, incidents, and concepts. Respondents comment on silence occurrences based on what they understand to be the case on the "inside of the evolving situation" (Dekker 2002). Participants in flight situations conduct their flights "as thousands they have flown before, until the moment when things start to go wrong" (Dismukes et al. 2007). Even in case of accidents, the crews and the controllers could not anticipate the outcome of such flights, while operating for them.

There are several critical phases in cockpit-air traffic controller communication cycle or continuum, and silence is part of the interaction. The start is the briefing on the ground, the preflight ritual of walking around the aircraft till communicating the banner of "seat belt sign" off, and communicating with intermediate control towers. When approaching the destination, topical awareness is exercised through interaction with co-pilot and air traffic controllers. That means that silence flows in the conversation patterns used during flight communication and has an interactional effect for the work of both pilots and air traffic controllers. Upon arrival in destination, a debriefing is an integral part especially in military aviation. The debriefing is a meta-communication process which retracts the whole flight, provides a mechanism to evaluate human error and safety culture in a systemic way, and cultivates collegiality and leadership.

Two flights constructed a system of reference regarding silence for all responders who were on their duty when the accidents occurred. Silence was a latent phenomenon before takeoff and, in the landmark of the Falcon "VIP flight," had a deeper content in organizational reporting. Breaking this silence connects with the increasing trend of scapegoating for human error which has consequences in aviation safety. Silence was a dominant phenomenon in the Helios 522 "ghost flight" and redefined what it means to communicate in a timely manner, without waiting and remaining silent for feedback.

Air traffic controllers brought up a categorization which prioritizes in-flight activities and no-talk periods with distress or emergency situations. But there are also instances when silence is the magic word to keep. The magic of silence is observed when ATC is the receiver of the message, when it modifies the content of cockpit communication, and when it is required in the manuals and in intervals during managing an emergency. Sometimes silence is the result of an informal grapevine microculture of humiliation which interferes with safety culture of the airliner.

Moving on to fighter pilots culture and expertise, mutual silence is described with the implicit personality in a relation of closeness and mentoring which may have a downside. Silence could be a result of lack of situation awareness or could also cause incomplete situation awareness. Also, silence could create operational barriers in checklist automation, could challenge the chain of command, or could be a result of a perceived conflict between efficiency/passenger comfort and safety.

Silence is taking turns with interaction in emergency situations to protect the peace of mind necessary to deal with the mission critical environment, human lives, and integrity of equipment. Thus, aviators and controllers agree to proceed or take the next turn from a period of silence, when they want to receive/perceive something that pilots say, something that pilots say or do, or something about how or why they say or do it. That is why silence needs to complement itself with vocal communication. In the next chapter voice succeeds silence.

References

Asian Disaster Preparedness Center (ADPC). (2006). *Safer cities 18: The boy who cried, "Wolf!" or why a community-based alert system is a good idea*. Klong Luang Pathumthani: ADPC-USAID/ODPA Case Studies.

Alexander Hamilton Institute (AHI). (2002). *Whistleblowing: A new perspective on an old issue*. Ramsey.

Allison, T. G. (2017). *Destined for war: Can America and China escape Thucydides's trap?* Boston: Houghton Mifflin Harcourt.

Ashman, A., & Telfer, R. (1983). Personality profiles of pilots. *Aviation, Space and Environmental Medicine, 54*(10), 940–943.

Bach, G. R., & Wyden, P. (1968). *The intimate enemy*. New York: Avon.

Barrett, R. (1982). The human equation in operating a nuclear-power plant. In D. Sills, C. Wolf, & V. Shelanski (Eds.), *Accident at Three Mile Island: The human dimension* (pp. 161–171). Boulder: Westview Press.

Barshi, I. (1997). *The effects of linguistic properties and message length on misunderstandings in aviation communication* (Unpublished Ph.D. Dissertation in Linguistics). University of Colorado, Boulder.

Bennis, W. G., & Nanus, B. (2003). *Leaders: Strategies for taking charge*. New York: Harper Business.

Berry, J. W., Poortinga, Y. H., Segall, M. H., & Dasen, P. R. (1992). *Cross-cultural psychology: Research and applications*. Cambridge: Cambridge University Press.

Bienefeld, N., & Grote, G. (2012). Silence that may kill when aircrew members don't speak up and why. *Aviation Psychology and Applied Human Factors, 2*(1), 1–10.

Bradac, J. J., Bowers, J. W., & Courtwright, J. A. (1979). Three language variables in communication research: Intensity, immediacy and diversity. *Human Communication Research, 5*(3), 257–269.

Bruhn, J. (2005). *The sociology of community connections*. New York: Springer.

Burgoon, J., Guerrero, L. K., & Floyd, K. (2010). *Nonverbal communication*. Boston: Allyn and Bacon.

Chalmers, L., & Keown, P. (2006). Communities of practice and professional development. *International Journal of Lifelong Education, 25*(2), 139–156.
Christakis, N. A., & Fowler, J. H. (2009). *Connected: The surprising power of our social networks and how they shape our lives*. Boston: Little, Brown & Company.
Cookson, S. (2009). Zagreb and Tenerife: Airline accidents involving linguistic factors. *Australian Review of Applied Linguistics, 32*(3), 22.1–22.14.
Crampton, S. M., Hodge, J. W., & Mishra, J. M. (1998). The informal communication network: Factors influencing grapevine activity. *Public Personnel Management, 27*(4), 569–584.
Cushing, S. (1994). *Fatal words: Communication clashes and aircraft crashes*. Chicago: The University of Chicago Press.
Degani, A., & Wiener, E. L. (1993). Cockpit checklists: Concepts, design, and use. *Human Factors, 35*(2), 345–359.
Dekker, S. W. A. (2002). Reconstructing human contributions to accidents: the new view on error and performance. *Journal of Safety Research, 33*, 371–385.
Dekker, S. W. A. (2011). Systems thinking 1.0 and systems thinking 2.0: Complexity science and a new conception of "cause". *Aviation in Focus: An International Aeronautical Journal, 2*(2), 21–39.
DeVito, J. A. (2012). *Human communication the basic course* (12th ed.). Boston: Allyn and Bacon.
Dismukes, R. K., & Berman, B. (2010). *Checklists and monitoring in the cockpit: Why crucial defenses sometimes fail* (NASA Technical Memorandum (NASA TM-2010-216396)). Moffett Field: NASA Ames Research Center.
Dismukes, R. K., Berman, B. A., & Loukopoulos, L. D. (2007). *The limits of expertise: Rethinking pilot error and the causes of airline accidents*. London: Ashgate Publishing.
Doak, S., & Assimakopoulos, D. (2010). Tacit knowledge: A needed addition to SOPs, in a forensic science environment. *Forensic Science Policy & Management: An International Journal, 1*(4), 171–177.
Dominguez, C., Vidulich, M., Vogel, E., & McMillan, G. (1994). *Situation awareness: Papers and annotated bibliography*. Armstrong Laboratory, Human System Center, ref. AL/CF-TR-1994-0085.
Eden, D. (1992). Leadership and expectations: Pygmalion effects and other self-fulfilling prophecies in organizations. *The Leadership Quarterly, 3*, 271–305.
Ehrenhaus, P. (1988). Silence and symbolic expression. *Communication Monographs, 55*(1), 41–57.
Endsley, M., & Jones, D. (2012). *Designing for situation awareness: An approach to user-centered design* (2nd ed.). London: Taylor and Francis.
Foushee, H. C., & Helmreich, R. L. (1988). Group interaction and flight crew performance. In E. L. Wiener & D. C. Nagel (Eds.), *Human factors in aviation* (pp. 189–227). San Diego: Academic Press.
Fry, G. E., & Reinhardt, R. F. (1969). Personality characteristics of jet pilots as measured by the Edwards personal preference schedule. *Aerospace Medicine, 40*(5), 484–486.
Fuller, E. (2011). *You can't lead with your feet on the desk*. Hoboken: Wiley & Sons.
Gaba, D. M., Singer, S. J., Sinaiko, A. D., Bowen, J. D., & Ciavarelli, A. P. (2003). Differences in safety climate between hospital personnel and naval aviators. *Human Factors, 45*(2), 173–185.
Gladwell, M. (2011). *Outliers: The story of success*. New York: Little, Brown & Company.
Glassner, B. (1999). *The culture of fear: Why Americans are afraid of the wrong things*. New York: Basic Books.
Gordon, T. (1975). *PET.: Parent effectiveness training*. New York: New American Library.
Grote, G. (2014). Promoting safety by increasing uncertainty – Implications for risk management. *Safety Science*. Available online 1 March 2014, ISSN 0925-7535, https://doi.org/10.1016/j.ssci.2014.02.010.
Habermas, J. (1991). *The structural transformation of the public sphere*. Cambridge: MIT Press.
Habermas, J. (1998). *Between facts and norms*. Cambridge: MIT Press.
Hackman, M. Z., & Johnson, C. E. (1991). *Leadership: A communication perspective*. Prospect Heights: Waveland Press.

Handelman, D. (1990). *Models and mirrors: Towards an anthropology of public events*. Cambridge: Cambridge University Press.

Hayward, B. (1997). *Culture, CRM and aviation safety*. ANZSASI, Asia Pacific Regional Air Safety Seminar, Brisbane.

Hofstede, G. J. (1980). *Culture's consequences: International differences in work related values*. Beverly Hills: Sage.

Hofstede, G. J. (2009). Research on cultures: How to use it in training? *European Journal of Cross-Cultural Competence and Management, 1*(1), 14–21.

Hofstede, G., & Hofstede, G. J. (2005). *Culture and organizations: Software of the mind* (2nd ed.). New York: McGraw-Hill.

Hwang, W. Y., Hsu, J. L., Tretiakov, A., Chou, H. W., & Lee, C. Y. (2009). Intra-action, interaction and outeraction in blended learning environments. *Educational Technology & Society, 12*(2), 222–239.

Jacovides, A. (2012). *Delimitation practice in the eastern mediterranean* (Rhodes Academy of Maritime Law and Policy). Charlottesville: University of Virginia.

Jandt, F. E. (2004). *Intercultural communication* (4th ed.). Thousand Oaks: Sage.

Janis, I. (1983). *Victims of group thinking: A psychological study of foreign policy decisions and fiascoes* (2nd ed.). Boston: Houghton Mifflin.

Jaworski, A. (Ed.). (1997). *Silence: Interdisciplinary perspectives*. Berlin: Mouton de Gruyter.

Job, M. (1994). *Air disaster* (Vol. 1). Weston Ck: Aerospace Publications.

Johnson, S. D., & Bechler, C. (1998). Examining the relationship between listening effectiveness and leadership emergence: Perceptions, behaviors and recall. *Small Group Research, 29*(August), 452–471.

Kariotis, T. (Ed.). (1997). *Greece and the law of the sea*. Berlin/New York: Springer.

Katz, S. B. (1992). The ethic of expediency: Classical rhetoric, technology, and the Holocaust. *College English, 54*(3), 255–275.

Klein, S. M. (1992). Employee involvement and the communication process. In S. M. Klein & J. S. Niblock (Eds.), *Proceedings of the 1992 Conference of the Midwest Management Society* (pp. 111–14).

Kochan, J. (2009, December 10). *HCEP lecture: 'In search of the expert pilot- A research Odyssey in human factors'*. Drexel Human Cognition Enhancement Program (HCEP at http://www.drexel.edu/research/humancognition/). Online Video File at: http://media.irt.drexel.edu/mediasite/Viewer/?peid=8f7355a186f947798bd6bf7db224fefe. Accessed 6 May 2010.

Koestler, A. (1973). The tree and the candle. In W. Gray & N. D. Rizzo (Eds.), *Unity through diversity: A festschrift for Ludwig von Bertalanffy* (pp. 287–314). New York: Gordon and Breach.

Konstantaras, D. (2007). *24 Seconds (24 Defterolepta)*. Athens, Greece: Kastaniotis Publications.

Lesser, E. L., & Storck, J. (2001). Communities of practice and organizational performance. *The IBM Systems Journal, 40*(4), 831–841.

Leveson, N. (2002). A new accident model for rengineering safer systems. MIT Engineering Systems Division Internal Symposium, Boston: 1–27, MIT.

Leveson, N. (2012). *Engineering a safer world: Applying systems thinking to safety*. Cambridge, MA: MIT Press.

Leveson, N., de Villepin, M., Srinivasan, J., Daouk Neogi, N., Bachelder, E., Bellingham, J., Pilon, N., & Flynn, G. (2002). A safety and human-centered approach to developing new air traffic management tools. In *Proceedings of the 4th USA/Europe Air Traffic Management Seminar*, 3–7 December 2001, Santa Fe.

Lewin, K. (1947). *Human relations*. New York: Harper and Row.

Loukopoulos, L. D., Dismukes, R. K., & Barshi, I. (2001). Cockpit interruptions and distractions: A line observation study. In R. Jensen (Ed.), *Proceedings of the 11th international symposium on aviation psychology*. Columbus: Ohio State University Press.

Loukopoulos, L. D., Dismukes, R. K., & Barshi, I. (2003). Concurrent task demands in the cockpit: Challenges and vulnerabilites in routine flight operations. In *Proceedings of the 12th international symposium on aviation psychology* (pp. 737–742). Dayton: The Wright State University.

Loukopoulos, L. D., Dismukes, R. K., & Barshi, I. (2009). *The multitasking myth: Handling complexity in real-world operations*. Farnham: Ashgate.

Loyarte, E., & Rivera, O. (2007). Communities of practice: A model for their cultivation. *Journal of Knowledge Management, 11*(3), 67–77.

Merrin, A. C. (1996). *National culture and work attitudes in commercial aviation: A cross-cultural investigation* (Unpublished doctoral dissertation). The University of Texas at Austin.

Nardi, B. A., Whittaker, S., & Bradner, E. (2000). Interaction and outeraction: Instant messaging in action. In *Proceedings of the 2000 ACM Conference on CSCW* (pp. 79–88).

Nembhard, I. M., & Edmondson, A. C. (2006). Making it safe: The effects of leader inclusiveness and professional status on psychological safety and improvement efforts in health care teams. *Journal of Organizational Behavior, 27*, 941–966.

Oral, N. (2009). *Non-ratification of the 1982 LOS Convention: An Aegean dilemma of environmental and global consequence* (Vol. 1). Publicist (on-line) publication of the Berkeley Journal of International Law, The Regents of the University of California.

Orasanu, J. M., Fischer, U., McDonnell, L. K, Davison, J., Haars, K. E., Villeda, E., & VanAken, C. (1998). How do flight crews detect and prevent errors? Findings from a flight simulation study. In *Proceedings of the Human Factors and Ergonomics Society 42nd Annual Meeting* (pp. 191–195). Santa Monica: HFES.

Ostrom, E. (2005). *Understanding institutional diversity*. Princeton: Princeton University Press.

Pressfield, S. (2011). *The warrior ethos paperback*. New York: Black Irish Entertainment.

Ragins, B. R., & Kram, K. E. (Eds.). (2007). *The handbook of mentoring at work: Theory, research, and practice*. Thousand Oaks: Sage Publications.

Reinhardt, R. F. (1970). The outstanding jet pilot. *American Journal of Psychiatry, 127*(6), 732–736.

Retzlaff, P. D., & Gibertini, M. (1987). Air force pilot personality: Hard data on the right stuff. *Multivariate Behavioural Research, 22*, 383–389.

Rich, J. (1995). *War and society in the Greek world*. New York: Routledge.

Roberts, J. (2006). Limits to communities of practice. *Journal of Management Studies, 44*(3), 621–622.

Roitsch, P. A., Babcock, G. L., & Edmunds, W. W. (1978). *Human factors report on the Tenerife accident*. Washington, DC: Air Line Pilots Association.

Roulston, M. S., & Smith, A. (2004, April). The boy who cried wolf revisited: The impact of false alarm intolerance on cost–loss scenarios. *Weather and Forecasting, 19*. American Meteorological Society.

Schafer, M., & Crichlow, S. (2002). The process-outcome connection in foreign policy decision making: A quantitative study building on groupthink. *International Studies Quarterly, 46*, 45–68.

Schwappach, D. L. B., & Gehring, K. (2014). Trade-offs between voice and silence: A qualitative exploration of oncology staff's decisions to speak up about safety concerns. *BMC Health Services Research, 14*, 303.

Shuang, L., Gallois, C., & Volčič, Z. (2010). *Introducing intercultural communication: Global cultures and contexts*. London: Sage Publications.

Sturken, M. (1998). The wall, the screen, and the image: The Vietnam Veterans Memorial. In N. Mirzoeff (Ed.), *The visual cultural reader* (pp. 163–178). London: Routledge.

Triandis, H. C., & Suh, E. M. (2002). Cultural influences on personality. *Annual Review of Psychology, 53*, 133–160.

Tsolakis, A. D. (2013). *Dare! A way of life (Tolma! Enas Tropos Zois)*. Athens: 11Aviation Publications.

Turner, V. (1969). *The ritual process: Structure and anti-structure*. Ithaca: Cornell University Press.

Vakola, M., & Bouradas, D. (2005). Antecedents and consequences of organizational silence: An empirical investigation. *Employee Relations, 27*, 441–458.

Vasterman, P., Yzermans, C. J., & Dirkzwager, A. J. E. (2005). The role of the media and media hypes in the aftermath of disasters. *Epidemiology Review, 27*, 107–114.
Weinberg, H. L. (1958). *Level of knowing and existence*. New York: Harper & Row.
Wellman, B. (1983). Network analysis: Some basic principles. *Sociological Theory, 1*, 155–200.
Wellman, B. (Ed.). (1999). *Networks in the global village: Life in contemporary communities networks in the global village*. Boulder: Westview Press.
Wenger, E., McDermott, R., & Snyder, W. (2002). *Cultivating communities of practice: A guide to managing knowledge*. Cambridge, MA: Harvard Business School Press.
Wilson, D., & Sperber, D. (1988). Mood and the analysis of non-declarative sentences. In J. Dancy, J. Moravcsik, & C. Taylor (Eds.), *Human agency: Language, duty and value* (pp. 77–101). Stanford: Stanford University Press.
Woods, D., & Cook, R. (1999). Perspectives on human error: Hindsight biases and local rationality. In R. S. Durso et al. (Eds.), *Handbook of applied cognition* (pp. 141–171). New York: Wiley.

Chapter 6
The Voice as Knowledge Operator of Choice

The core communication-related category emerged from the GT approach of Chap. 4, with the labels of silence and voice (Sect. 4.3.3). In this data analysis chapter, voice is explained in its synthetic nature as knowledge operator and complementing silence. Voice consolidates hesitation, paralanguage, direct language, and native language codes.

6.1 Organizing the Common Knowledge on Voice

Chapter 6 has a slight different structure in comparison with Chap. 5, as it includes three scenarios regarding the voice category: hesitation in the nonverbal but vocal part, language, and familiarity relationship which, then, directs the study to the mother tongue issue. Communication forwards to nonverbal channels with notions manifested in the responders' replies, introducing paralanguage and hesitation as a medium of interaction (Sect. 6.2). Then the Helios 522 flight (from Chap. 3, Sect. 3.1.2) is analyzed in terms of nonverbal and verbal communication signals in conjunction with the pair of F-16s that were scrambled and insights from the accident investigation (Sect. 6.3).

The next Sect. 6.4 uses the hesitation scenario as developed in three parts in Sect. 3.2.2 (captain reports, captain acts, first officer's comments). Four different responders (two pilots and two controllers) follow with detailed commenting on their communication concepts and attributes (Sects. 6.4.1, 6.4.2, 6.4.3, 6.4.4).

Then the scenario of language directness and familiarity is introduced, along with relevant communication concepts. The scenario develops in two parts (Sect. 6.5.1), and the responders argue on both parts (Sect. 6.5.2 with a controller, Sect. 6.5.3 with fighter pilot, and Sect. 6.5.4 with 6.5.4 with helicopter pilot). In the last part of the chapter (Sect. 6.6), the concept of mother tongue is presented, in the context of language and communication in aviation: "Is it easier to know what

T. Katerinakis, *The Social Construction of Knowledge in Mission-Critical Environments*, Innovation, Technology, and Knowledge Management,
https://doi.org/10.1007/978-3-319-91014-7_6

happens in your mother tongue or other?" Selected responses are presented to show converging topics, in language- thought- cognition. The concluding remarks (Sect. 6.7) summarize the major topics and introduce several concepts that extend the current book.

In this chapter, like in Chap. 5, I selected the responders who were able to connect the scenarios with insights from their personal experiences, when and while doing their job. From the complete list of Sect. 2.5 (Fig. 2.3), the subset participants (with standardized names) are shown below:

As we can see in Table 6.1, 11 participants with different profiles and backgrounds (military and civil aviation pilots, controllers, and an accident investigator) were selected for the depth of their insightful responses. With their

Table 6.1 Selected responders for the scenarios of "voice" narrative

1.	Eleftherios Venizelos, fighter pilot –AIP1	
2.	Ippokratis Koos, helicopter pilot – AIP6	
3.	Dimokritos Alexandroupolis, fighter pilot – AIP7	
4.	Ifaistos Limnios, civil aviation pilot-– AIP9	
5.	Diagoras Rodios, civil aviation pilot – AIP10	
6.	Megas Alexandros, accident investigator – AIP13	
7.	Ikaria Ikaros, ATC – AIP14	
8.	Oia Santorinis, ATC – AI1P5	
9.	Neos Aghialos, fighter pilot – AIP23	
10.	Palios Ellinikos, civil aviation pilot – AIP24	
11.	Stan Dardman, civil aviation pilot – AIP25	

interpretations we complete the conceptualization of the transition from silence to vocal messaging, as it happens while respondents exercise their roles. More details for these responders are included in Appendix E.

Analogically to Chap. 5, this chapter connects the narrative with concepts of communication theory, concepts of occupational and organizational culture, and concepts of group work, with the respective literature documenting the findings. This style matches the discourse analytic approach better, enriches the contributions of this book, and extends the philosophy of a GT approach for knowledge representation of flight realities by putting more emphasis in the raw experiences.

6.2 Paralanguage and Knowledgeable Communicative Transactions

This section organizes the background of paralanguage and hesitation talk in aviation communication as captured in the informal discussions with the responders of this study, prior to their replies to the specific scenarios.

The communicative direction of this study considers the cockpit (with the co-present controller included) as the kernel which transmits messages (non-talks are included as silence periods in an integrated flow of interaction). Communication behaviors involve verbal messages, gestures, vocalizations, other vocal expressions, and their combinations which develop in packages of microstructured momentary communicative events, like those observed in aviation where seconds and instances correspond to instrumental time for the flight situation. All parts of the system of messages work together, similarly to the information richness analyzed in the pioneering psycho-anthropological work of Pittenger (Pittenger et al. 1960).

Since no visual contact is feasible in the cockpit-controller interaction, the use of paralanguage, i.e., the vocal but nonverbal aspect of communication, is channeled through the same verbal contacts (Argyle 1988). Therefore, aviation communication uses voice, its sound, echo, and surrounding noises to carry verbal and nonverbal signals, not only in real time but also retrospectively, via the recording conversation and noises (and flight parameters) stored in FDR and CVR recording/indication monitoring devices (explained in Sects. 5.2 and 5.5.2).

When comparing routine cockpit interaction and crisis interactions, significant differences in meaning are communicated depending on where the speaker put stress in the message. Also, paralanguage cues include the pitch, voice qualities like volume and rhythm, pauses and hesitation, and vocalic while crying, whispering, yelling, etc. (Trager 1958). Paralanguage cues, in certain contexts, are an indication of an emotional state or personality, especially if the sender and receiver speak the same language or share a mother tongue. And since emotions are not just reflections but also include the *cognitive representation of behavioral reactions*, they relate to the evolution of communication patterns (Salem 2009). That is why aviation communication is rich in the use of voice mechanisms (including language and paralanguage) which are part of sense-making process when low-context language is the medium to convey the standard of the routine, the pressure of the emergent, the agony of life risk, or the moral judgment in applying fighting rules of engagement or aviate in free-fire zones when the pilot has the sole authorization for weaponry use.[1]

In-flight interactions follow the conventional principle of transactional communication about the elements and signals in message content/style (Devito 2011); i.e., the transactional nature includes (i) communication is an ongoing activity with its elements in constant change; (ii) a source is there only with a receiver and no message exists without a source (interdependency exists); (iii) communication is influenced by the individuals' interpretations, although they have a strong standardized foundation; and (iv) every participant is both a sender and a receiver. So, speaking, listening, and refusal to communicate is itself a mode of communication.

6.3 Knowledgeable Cockpit Verbal and Nonverbal Communicative Actions

Megas Alexandros (introduced in Chaps. 2 and 3), one of the world's most experienced accident investigators and discussant for this study, underlines the communicative nature of the flying activity of the agonizing fighter pilots who were scrambled,[2] in the Renegade process, for Helios 522 "ghost plane" flight (of Sect. 3.1.2). Observing in more detail in the encounters of the F-16s pair with the "ghost plane," the study identifies the manifestation of nonverbal communication (aircraft-to-aircraft and pilot-to-aviating purser communication), the emotional parameters communicated in the Renegade process, and the verbal-nonverbal communication mix in the declaration of emergency from two different channels. These detailed elements formulate the "what and how" we know about the flight.

Megas Alexandros visited the two F-16 pilots in their home Air Force Base (111CW) to discuss their experience instance by instance algorithmically. Due to the accident investigation, their discussion took the form of a detail account (in protocol analysis terms of Sect. 3.2). This approach is described in the accident

[1] The descriptive narrative of the interplay of emotions, actions, improvisations, and operational decisions from Lt. Heather "Lucky" Penney, is a picturesque example of such communication patterns. She was one of the first two combat pilots in the air from Andrews Air Force Base in the morning of September 11, 2001, after the hijacked plane attacks in New York World Trade Center. Her recollections and sharing of the developing situation awareness, different styles of communication, small-talk, personal relations and trust even in the higher offices are compatible with the current analysis (despite of some criticism she has received). See more at the video interview http://www.c-span.org/video/?300959-1/major-heather-penney-september-11-2001 (August 11, 2011).

[2] Scrambling aircrafts means to get them into the air as quickly and as possible and into formation to defend, interdict, intercept, etc. In HAF quick response alerts, 2 and 5 min are typical time frames. Quite often when in scramble, there is no time for mission briefing, but just the order and rules of engagement are exchanged in the life vests room. Originally, during WWII in the Battle of Britain, scrambling aircraft was referring to immediate takeoff in order to avoid a catastrophic attack against them in the ground.

report (HAAISB 2006) and a special book (Tsolakis 2013). I asked Megas Alexandros to elaborate further in the report's description so that I could trace the communication signals. I have also used the extended testimonies during several judicial procedures in Cyprus and Greece, as well as reports publicized by the HAF's General Staff regarding the interception phase.[3]

Helios 522 and F-16s Parallel Flying The original pilot and copilot of the Helios flight were incapacitated and fell unconscious. The autopilot was engaged with the programming to the destination airport of Athens. The autopilot[4] led the aircraft to predetermined patterns around the Athens Area Control Center, the primary flight dispatcher in Greece's capital. What did the leading intercepting pilot, namely, *Neos Aghialos*, do next to communicate? The passage below comes from the discussion with Neos Aghialos (with explanatory text in parentheses added):

> When the two F-16[5]s flew around Helios flight in loops and maneuvering dangerously close with shaking movements and oscillations (considered "aircraft body movements") as prescribed in the interception protocol to "gain attention" for almost 12 minutes but no signal received from the Helios flight (recorded also in the F-16 aircraft camera). The leading pilot for the pair of F-16's (flying side-by-side to Helios) continued with hand gestures (as radio communication was with no reply[6]) for which he perceived feedback visually by the purser who (with limited oxygen) was attempting to regain control of the aircraft from the autopilot.[7]

The two F-16s flew around the Helios plane in a reconnaissance mission to obtain information by visual observation and "attract" any signals to and from the Helios

---

[3]The Helios 522 flights crashed in Greece's mainland, so the relevant judicial took place in the company's headquarters in Cyprus and in Greece where the tragic accident actually happened. For this study, I was able to use the judicial proceedings (from Cyprus and Greece) that are part of State Archives, after the last phase of the judicial processes in the Supreme Court on February of 2013.

[4]Eurocontrol's skybrary explains that: "An autopilot is a device used to guide an aircraft without direct assistance from the pilot. Early autopilots were only able to maintain a constant heading and altitude, but modern autopilots are capable of controlling every part of the flight envelope from take-off to landing. Modern autopilots are normally integrated with the flight management system (FMS). The FMS is an on-board multipurpose navigation, performance, and aircraft operations computer designed to provide virtual data and operational harmony between closed and open elements associated with a flight from pre-engine start and takeoff, to landing and engine shut-down." (see at http://www.skybrary.aero/index.php/Autopilot and http://www.skybrary.aero/index.php/Flight_Management_System).

[5]As the manufacturer, Lockheed Martin Corporation explains "the F-16 Fighting Falcon has proven itself as the world's most capable 4th generation multi-role fighter, serving as the workhorse of the fighter fleet for 28 customers around the world, for more than 40 years with constant upgrades (4500 aircraft and 55 repeat buys have been supported)." More at http://www.lockheedmartin.com/us/products/f15.html.Scrambles are typical missions for F-16.

[6]The internationally established International Civil Aviation Distress Frequency to be used for Distress and Safety by Search and Rescue is 121.5 MHz (see http://www.skybrary.aero/index.php/Distress/Emergency_Frequencies).This flight was not responding despite the attempts as it was committed to the frequency of the departing airport, as Megas Alexandros noted.

[7]DNA testing of genetic material of the cockpit debris and oxygen masks after the crash appear to confirm the presence of the purser later identified as Andreas Prodromou in the cockpit (HAAISB 2006; p.69; Tsolakis 2013, p.394-395).

flight. The interception is a second step, which theoretically may escalate till a "shooting down." Shooting down is ordered when there is a threat for a populated area or critical national infrastructure on the ground that cannot be excluded as risk, in case an aircraft (its crew or passengers) does not signal any response or any other plausible/legible message.

Neos Aghialos continued from his fighter jet wing gestures to his own hand gestures in order to "elicit" a signal of understanding. So, the two F-16s took off as a dyad with Neos Aghialos, as the pair leader and the other F-16 in support and targeted flight which would radar-lock in the tail of the Helios plane. The F-16's radar searches and tracks the sky in a zigzag scheme and the radio beam reflected by Helios plane; Helios indicator was shown on radar display and circled around to lock for weapon routing, if decided according to the Rules of Engagement of the Renegade response protocol.

The F-16 radar tool functions in a mechanical situation awareness mode towards the intended target as well as the visual display and affords the operational capacity for the pilot (Endsley 2004; Dominguez et al. 1994); the pilot operator has the training to rely on the automated radar tool and the judgment to avoid misuse and disuse of automation, following the terms of Lee and See (2004). Since the time of their scramble, the pair of F-16s is in a monochronic time orientation as they observe the situation with a schedule of one thing at a time, and the next flight action (on time slot scheduled) depends on the outcome (and response from Helios and from their commanding centers) of the current one; it is an instance of a chronemic communicative dimension in the microenvironment they were flying as described by Hall (1963).

Their operational clock has several measures of time. Helios airplane was flying almost 3 h running out of fuel; the two F-16s were flying with different maneuvers (the second in a more stable and conservative way), in order to cover different visual angles and save fuel. The replacement dyad of Mirages 2000–5 was in readiness to scramble and continue the Renegade interception if needed. The time for the Renegade shootdown was counting down in the Greek Pentagon, where the Defense Minister and the Chief of HAF was in direct line with the Prime Minister. The survival time of the person in Helios pilot seat was also limited (since depending on the supply of oxygen which was running out, even the F-16 camera recorder overflowed in storage capacity during the flight time).

All these flight time determining indicators were counting down as sand glasses, for the F-16 pair, as if they were having a mission briefing while flying. That was the description in original military classified (and then released) debriefing reports to the HAF General Staff, prior to the start of HAAISB accident investigation and are not included in the report. Of course, this realization of time restrictions communicates a volatile situation awareness state, for both pilots and for Neos Aghialos in particular, in a dynamic microenvironment of continuous adaptation.

Neos Aghialos' approach was a practice of an unusual artifactual communication using the plane as a communication tool (by virtually waiving the F-16 wings as communication illustrators) and an emblem in kinesics (Knapp and Hall 2010) in nonverbal communication which invalidated the initial working hypothesis of a hijacking. The nonverbal interaction of signals (waiving hands with visual contact) formulated a short verbal message from Neos Aghialos in cooperation with the information to Athens Area Control (Approach) Center and the Chief of the Air Force (see below, with explanatory text in parenthesis added):

> The man in the cockpit shows signs that he has seen me… Now I can distinguish a second person sitting the co-pilot seat (with flight attendant's attire) and pointing down. I can see that person in the pilot's seat is trying to hold the controls and stabilize the aircraft. There is an open book (like a manual0 beside him.

Neos Aghialos tries a glance before attempting contact and conveys a message with positive interpretation (of no attack, just before the crash) by forwarding verbally the nonverbal reactions he has noticed, after his eye contact (from the F-16 cockpit to the Helios cockpit) with the person on the captain's seat. Neos Aghialos and the person in the captain's seat were two physically distributed pilots-in-command with a sense of copresence in their parallel routes. In the cockpit-by-cockpit communication (from the Helios to the F-16), the proximity of the two aircrafts communicates immediacy of action, similarly to the proxemics dimension via the visual channel (Hall 1963), and seems like a manifestation of one of the facial feedback hypotheses (McIntosh 1996); the facial expression action of the struggling Helios purser with the flight manual had an impact to the physiological arousal of Neos Aghialos. Neos Aghialos was able to conceive the pressure and determination of the person who was making the effort to disengage the autopilot, use radio communication, and aviate the Helios plane (when all the others suffered from hypoxia and were incapacitated).

The person in the pilot seat of Helios flight transmitted a signal of his desperate effort to regain control of the flight, although most likely unintentionally, when he used the assistance of the manual. As Neos Aghialos spotted the manual, it became a central marker reserving the cockpit as territory, in the way territoriality communicates possessiveness (Moore et al. 2010). Therefore, a relief sign was the feeling for the Athens Area Control Center and the National Center of Air Force Operations (and other military vigilance centers) when they realize that there was no eminent terrorist threat and the risk of out-of-flight civilian casualties was reduced, when the plane was directed out of the densely populated neighborhoods and public beaches in the seaside area of the capital city of Athens. The tragedy which was about to happen would affect everyone on the flight, but not others, so that the number of casualties would be as reduced as possible.

Emergency Calls The next point that connects the Helios ghost flight with verbal communication is that two distinct sequential messages for an emergency (the crash which was approaching) were declared for that flight, in completely different channels. The first one was spoken by the person in the captain's seat but not transmitted outside the cockpit (comments added, HAAISB 2006):

> CVR Time: 08 54 18"
> MAYDAY, MAYDAY, MAYDAY, HELIOS AIRWAYS
> FLIGHT 522 ATHENS (…unintelligible word)
> CVR Time: 08 55 05"
> MAYDAY (very weak)
> CVR Time: 08 55 09"
> MAYDAY (very weak)

How could we know that? Is anyone listening? It is an excerpt from the transcript of the flight as recorded (internalized) in the CDR. The message is an explicit declaration of an emergency during the initial descent of the flight. But the microphone key was not pressed for the VHF radio[8] to transmit. Thus, the message being sent did not reach the intended target/receiver, as no appropriate channel was used and no one was listening (outside Helios plane). The person-sender was close to oxygen starvation losing his breath and with poor situation awareness due to his physiological condition. The stress and pitch in the tone of voice relays the message of the critical moment of the last surviving cabin crew member who was seating in the captain's seat. This message does not have any informational content till the moment of the crash and is now acquired as hindsight for the accident investigation.

The other emergency message about the Helios flight was transmitted from the outside of the flight, from Neos Aghialos the leader of the two F-16s which was flying cockpit-to-cockpit, side-by-side of the Helios flight (as close as 15 m, around 50 feet; it was a "breath's distance" in their culture as the two pilots pointed out). Neos Aghialos, in his agonizing effort to communicate (even if he would just receive a signal), has approached a proxemic distance of face-to-face interpersonal communication as described by Hall (1966/1990). He flew beyond the vertical and horizontal separation standards to facilitate the attempt to clarify the situation, with any mode of communication he had available. The Helios flight crashed with an impact on hilly terrain at 12:04:32 p.m. (August 14, 2005), and the emergency message signals were explicit. The leading F-16 pilot Neos Aghialos declared the emergency (as recorded in the video archive of the F-16 flight camera), for the other aircraft (explanatory text, in parenthesis, added):

> Mayday-Mayday-Mayday: the civil aviation plane crashed on top of a mountain
> Mayday-Mayday-Mayday: the civil aviation plane crashed on top of a mountain
> Stanby, Stanby: I will give you the coordinates
> Standby, Standby: the airplane crashed outside any residential area
> *(Continues with a pause for silence of about a minute to breathe and then a repetition and confirmation of the geographic coordinates of the crash location)*

[8]VHF, stands for very high frequency, is conventionally defined as the portion of the electromagnetic spectrum of frequencies between 300 and 30 megahertz including any radiation wavelength between 1 and 10 meters. VHF signals are usually used for television and radio transmissions. As Eurocontrol skybrary explains: "Aircraft Communications, Addressing and Reporting System (ACARS) is a digital data link system for the transmission of messages between aircraft and ground stations, which has been in use since 1978. At first it relied exclusively on VHF channels but more recently, alternative means of data transmission have been added which have greatly enhanced its geographical coverage. There has also been a rapid trend towards the integration of aircraft systems with the ACARS link. Both have led to rapid growth in its use as an operational communications tool" (See more at http://www.skybrary.aero/index.php/Aircraft_Communications,_Addressing_and_Reporting_System). Communication frequency is crucial because it restricts the channel of signals which should be open to all airport traffic. In ICAO Doc, 9870, "Manual on the Prevention of Runway Incursions," the provision explains: 4.2.6 "All communications associated with the operation of each runway (vehicles, crossing aircraft, etc.) should be conducted on the same frequency as utilized for the take-off and landing of aircraft." Further details are provided in Appendix A, §1.5. (See at http://skybrary.aero/index.php/Use_of_aerodrome_Tower_VHF_frequency_by_vehicle_drivers_involved_in_runway_operations).

It was a feed-forward message in a continuum (for almost 5 min) until the moment of the emergency declaration, from the F-16 pilot, for the irreversible route to crash (the main tragic "event and message"). Although the message was explicit and was transmitted and received from various control centers, the intensity of the moment to become aware as an eye witness of not just another crash but one hundred and twenty-one people deaths (most likely at the moment) was compacted in this moment of breathing. It was an unrepeatable communication moment for all receivers in control centers who perceived and decoded the emotional state of the pilot (Burns and Beier 1973; Beier 1974).

In general, fighter pilots are trained in emotional discipline, self-awareness, and control and are tested periodically, physiologically, and psychologically; but instant emotional gales for incoming stimuli are encouraged as a blowoff externalization process (at least in the culture of HAF). Parenthetically, when I brought the topic of the emotional stability of a fighter pilot and their mission for an informal comment during discussions, *Dimokritos Alexandroupolis*, a Major General in the HAF, commented on the culture of self-preservation through the group of pilot colleagues (explanatory text in parentheses added):

> A fighter pilot leaves every day for his[9] base to takeoff (in real-life defense drills in flights) without knowing whether he is coming back or not. When a colleague pilot dies on duty or drill a teardrop comes in the eye till the formation lands-back in the base. We learn it the hard way in the Air Force Academy, not because we are trained to, but because we realize it on practice: every year we call-over the classmates who did not survive the flight training. Then life goes on and airmanship spirit is cultivated.

The value of human life and the collective orientation and the strong sense of belonging in the group (even post-mortem) transcend individual achievements and personal ambitions among fighter pilot who were classmates or fly together (even for a short period of time). Small interconnected pilot networks formulate by starting from dyads who fly together and who participate in the same briefing and missions and then in the same quadruplets (and so on, in single- or two-seat fighter planes). In this sense microcultures develop, and the bonding becomes tight. Death rituals are enforcing airmanship culture and awareness. I have experienced such instances during my military duty in 115 CW/Base operations Directorate, in Greece, when two different pilots who I have scheduled to fly in two different regular training days did not return. Silence was shadowing the base till the time their formations returned. In the next day, I had to participate in the honoring squad, as the Tactical Chief of the Air Force (who supervises operations) was there to participate in the mourning ritual in an expected leadership act from his side.

---

[9] Although without restrictions in the past years, female fighter pilots are in the margin of 5% in HAF. The intensity of training and lifestyle, the family-oriented society, and the national exam competition to enroll to the Hellenic Air Force Academy are (among others) factors that have contributed to such numbers. Thus, the use of grammatical masculine gender pronouns, for fighter pilots, is not considered discriminatory. It is indicative that the popular drama series of the Greek Mega TV "Silence in the Air" (Anosis SA, 2000) featured two airwomen characters in military duty: an air traffic controller and the chief engineer.

A Long Closing Statement Coming back to Neos Aghialos' actions after his verbalization for terminal emergency declaration, there was no doubt for the unfolding tragedy, and the search and rescue response alert services were energized by HCAA. Instead of just verbalizing the message aloud, the pilot's inhale-exhale rhythm coupled with the falter (for some seconds), and pause was a clear closing statement. It was his own moment of silence, mourning for the people who crashed in Helios flight, and with a transient self-blame thought that he could not help them, although the Renegade mission was accomplished. The cockpit-to-cockpit flight of two co-present pilots on a captain's seat (in parallel flying) became their bond, taken into account what Dimokritos Alexandroupolis stated above. Silence and paralanguage were eloquent in transmitting the message.

Neos Aghialos commented, in the same discussion with Megas Alexandros for a dilemma from the other side of the coin, i.e., his Renegade mission: "I cannot forget the flight I was flying side-by-side, as I was seeing everything, communicating with the person in eye-contact but could not do anything to assist him practically and save them. Sometimes I am thinking aloud in the night, would it be possible to intercept and shoot down the plane with all those people, in case I was not able to communicate at all?" Self-efficacy as belief in personal abilities, aspects of fear, and trust in effort and command are enduring balancing attributes in the pilot culture. Still, the decision-making function is a floating inquiry, at a microlevel of what pilots evaluate with their sense till the tactical level of operation and the macro-level of public safety policies like the Renegade Protocol. Lastly, HAAISB committee was already (before the actual impact) approaching the crash site to Grammatiko area outside Athens (Greece) since they were engaged in the likelihood of a major accident from the early stages of the Renegade response process. Neos Aghialos' interception and messages were a feed-forward call for them for the "future" accident (to come) before it occurred.

The silent Helios "ghost" plane communicated to an additional recipient through the co-present F-16 pilot Neos Aghialos. The hilltop in Grammatiko area (in the suburbs of the Greek capital city of Athens) where the crash took place is known as "Black Mountain," a preexisted name which became symbolic of the catastrophe and the subsequent heavy particles and other chemicals scattered around and now under special environmental treatment (Tsolakis 2013). The trace of the debris and the semiotic content in the Black Mountain still communicates the haunted story.

6.4 Hesitation: Scenario and Knowledge Framework

The vocal but (still) nonverbal dimension of speech channel of communication has to do with how one says something and not with what one says. Among the voice qualities, prior to actual talking the language, pause and hesitation are perceived and understood in the broader verbal interactions. Adjustment is necessary for both sender and receiver ends to communicate. That is why a second scenario from an actual reported flight situation in order to investigate the transition from silence to language, via paralanguage.

In this scenario (laid in three parts as explained in Sect. 3.2.2), we move from nonverbal instances of hesitation as an indication of bypassing in cockpit-controller interaction or as a contributing factor of aviation accidents. Let us summarize how the scenario unfolds with the captain of the anonymous report, his actions, and the reaction of the first officer.[10] The quote follows:

> Captain: On takeoff the Tower Controller called us and sounded as if he had something to tell us, but did not know what to say. We both noted a tone of concern and hesitation, in his voice as if he was still unsure of something at that moment. We had to make an immediate decision.

The civil aviation pilot (the captain) reports that he has ended the taxiway movement, continued to the runway to take off, and started rolling to develop the necessary speed. The call from the air traffic controller was made with a pause and murmuring the call sign (which identifies the aircraft in their interactions; it is usually the carrier name with a number). Although the message was incomplete, the tone was thought-provoking, and timing was critical as they have started the speeding up for the takeoff process. They still had some route to go, and they had to decide with no other input, just their airplane data (fuel to avoid fire risk and the remaining distance to be safe). What did the captain do (in summary, below)?

> The Captain continues: I elected to initiate rejected takeoff procedures. During deceleration the Tower Controller said, "Disregard." The sound of one's voice, the tone and force, all convey a message. I did not like the message I was receiving and could not gamble that he was trying, but unable, to warn us of something ahead. I would take the same action again.

The captain decided to take the incomplete call as a reason to reject the takeoff and started the deceleration in order to stop the aircraft. At that point the call from the controller completed the message of hesitation with a *disregard*, as it was nothing to talk about, to ignore the call. Uncertainty was something he could not evaluate, in case it was a warning for a potential problematic situation ahead (they were cleared to take off and the call would project a reason to revoke that clearance). In the original Callback report, the first officer who was able to receive the same incomplete call from the controller comments below. What is their evaluation (in summary, below)?

> The First Officer's reports: I believe the rejected takeoff was the right thing to do. When you get a call from Tower at that point in the takeoff roll, the first thing that pops into your mind is "something's wrong." So, the Captain did what was prudent and safe by rejecting.

The first officer is fully confident for the action of the captain. It is a reaction consistent with the spiral of silence theory in communication, in which voicing agreement is more likely than disagreement in a controversial issue after evaluating potential consequences, as most respondents in this study agree for the choice of the captain (Windahl et al. 1992). It seems that, since the flight was actually accelerating to take off, a call from the tower breaks the sequence which seems to be uneventful. The potential of any unexpected event is a call for being prudent than being sorry.

---

[10] It comes from the NASA's ASRS Callback report, issue 395, January 2013, available at http://asrs.arc.nasa.gov/publications/callback/cb_395.html: situation from Boeing 737 Pilot's Report.

The hesitation scenario summarized above is the discussion topic for the discussants who shared their experience for this study. The inquiry was to evaluate the hesitation issue, to explain their action as if they were pilots, and to suggest ways to repair hesitation and other interference for the first officer. Sometimes, hesitation is also embedded in intra-plane communication between the cockpit and the cabin. Overemphasis in the sterile cockpit rule (the philosophy of which was explained in Sect. 1.3), unfixed team formation and separate briefing for the pilots and the cabin crew could result to ineffective communication between two crews (Chute and Wiener 1996).

6.4.1 Safety First with Clear Short Messages: Pilot's View

The start is the fighter pilot at the rank of Colonel of Hellenic Air Force, namely, *Eleftherios Venizelos*, who was presented in Sect. 5.5.1. Eleftherios Venizelos has active experience in national fighter missions in HAF, in international reconnaissance missions for NATO where he needed to take off and land in most of the 17 countries which consist the operational branch of NATO.[11] He is also informed about the incidents reported in the Flight Safety Directorate of the Hellenic Air Force General Staff, as well as for the NATO incidents. So, what is his opinion on the hesitation scenario?

> Since takeoff is a critical phase of flight, I find the performance of the controller very negative. He should have said a clear message using Standard Terminology without any hesitation. The crew had to take a quick decision depending on the speed, runway conditions and, runway remaining distance. There was no time to explain or to ask. I have never experienced something like this (Poor Communication in takeoff roll). If I were the pilot, I would have also rejected the takeoff. Firstly, because the speed was well below the critical V1 speed; secondly, because of the long runway remaining distance. So, safety first! Then, to explain to the passengers and to the company.

Eleftherios Venizelos is quite definite about the responsibility of the controller which is negatively characterized because of the unclear message and the incomplete standardization format of his vocalization. The crew approaches a point of no return (after a certain distance and acceleration to take off), so decision is immediate, and uncertainty is an obstacle. The unexpected call from the tower created a chain of thoughts for the value of orality as information, at this point critical (Turner 2010).

The incomplete sentence was as starter of a message with a high informational value, due to the fact that clearance is the terminating message and clearance was given to take off. The value, in risk terms, of the conditional "what tower could have said or mean" generates a potential safety threat which cannot be evaluated. Going

---

[11] North Atlantic Treaty Organization is an alliance founded after World War II. As quoted, at present, NATO has 29 members. In 1949, there were 12 founding members of the alliance: Belgium, Canada, Denmark, France, Iceland, Italy, Luxembourg, the Netherlands, Norway, Portugal, the UK, and the USA. The other member countries are Greece and Turkey (1952); Germany (1955); Spain (1982); the Czech Republic, Hungary, and Poland (1999); Bulgaria, Estonia, Latvia, Lithuania, Romania, Slovakia, and Slovenia (2004); Albania and Croatia (2009); and Montenegro (2017). For more information see https://www.nato.int/cps/en/natohq/topics_52044.htm?

from the oral start to the complete information which would be so important to be talked in the post-clearance phase, the pilot has to decide with a risk he cannot take. The cost of incomplete information accumulates to the cost for a safety-related risk for which there is no time for evaluation and control. Thus, almost automatically, the pilot's thought is "what it means to reject takeoff" in terms of technical feasibility: speed (V1 is the target speed after acceleration to take off), fuel, runway length, weight, and the institutional parameters of losing a clearance, bypassing tower's instruction (compliance challenge), and having to be accountable to passengers (marketing and public relations challenge) and the airliner itself (organizational culture challenge). As many challenges there may be, safety is the first in priority. What is Eleftherios Venizelos' suggestion, in his quote (capitals are used for emphasis)?

> There is a way to repair hesitation only if there is enough time for effective communication. I haven't trained specifically for this and there is no way to train crews for everything that could happen. On the other side all crews and ATC controllers know that takeoff is a critical phase and should avoid unnecessary messages. Clear short messages inside and outside the cockpit are necessary. That is the reason pilots use single words like "ABORT, REJECT, CONTINUE, etc". There is no time for the brain to analyze the situation and react. You hear this word, you execute the checklist immediately.

Eleftherios Venizelos has the pilot's point and more specifically the Air Force pilot who is trained in everyday duty to respond to flight inquiries, alerts, and scrambles in predetermined short time. Since time is limited and nonexistent when the airplane is on the move, repair is not possible (without a serious of turn-taking which starts from repetition). It is another indication of the clear affordances of time, in aviation (Gibson 1979); flights cannot be understood without the possibilities or impossibilities of action which time permits or constraints. Time is a parameter of the operating environment, a contextual element of the communication context of the flight which dictates some course of action like the rapid decision-making (here) or directs to certain restrictions of action. It is not an issue of specialized training, as training, in the end of the day, is a learning process of making decisions in unexpected circumstances and without (always) having practiced every single potential event or incident.

It is part of the sterile cockpit principle, as no interaction is expected in the post-clearance phase. Breaking that practice magnifies uncertainty as risk. A clear short message, in Eleftherios Venizelos' mind, with a verb in imperative mood (like *abort, reject, continue* which he emphatically capitalized) which would convey an absolute negation (like *discontinue, accept, stop*), would create the intended reference and avoid the problem of uncertain reference (in Cushing's terms, 1994). Analogically, short or one-word replies facilitate the confirmation messages from both sides.[12]

[12] In the flight US1549 (Accident Report, http://www.ntsb.gov/doclib/reports/2010/aar1003.pdf, Jan 15, 2009), a critical interaction was NY TRACON: "Alright cactus 1549. It's going to be a left. Traffic to runway 3–0.1" US1549 (replies): "Unable." A simple non-explanatory and clear-cut word conveyed the dramatic situation.

6.4.2 Pilot Interpretation and Controller's Bad Moment

In the same scenario of hesitation, the study incorporates the evaluation of an experienced civil aviation pilot, namely, *Diagoras Rodios*, who has personal involvement in quality assurance and control issues. He was responsible for the single-day transition of one of the two major Greek airliners from the old Athens Airport to the new "Athens International Airport Eleftherios Venizelos",[13] in Greece, during which he coordinated massive communication load. As one of the experts with a thorough consideration of aviation professionals (both pilots and controllers), he suggests the supremacy of the pilot's decisions, he is accommodating about verbal messages principles that affect meaning making, and explains why a take-off to continue could be possible.

> Obviously, it's a miscommunication in terms of controller's bad training, limited experience or just a "bad moment". In aviation, we never start transmitting a message unless we think about what we are going to say. I haven't experienced similar case, but from the description, the pilot must be found to a really difficult position during a critical phase of flight. Take-off cancellation is an uncommon event, at that stage. Quite in more rare occasions is due to harsh conditions (short runway, plane loaded till the margin, wet runway etc.). It is a pilot's decision which has to be justified to the airliner afterwards.

Diagoras Rodios seems to be more inclusive and open in his interpretation, as errors do occur regularly. It may be a bad moment and not a generic attribute of the controller's function. The aviation environment requires precision, accuracy, efficacy, and predictability. So, a message is transmitted only when it is complete, in order to be meaningful. Takeoff and landing are among the essentially busy phases of flight – taxi out, take off, initial climb, intermediate and final approach, landing, and taxi in. No post-clearance interaction is expected at those times. There are occasions when technical parameters or the physical environment demands a cancellation. As it affects instructions, schedules, and corporate brand, any cancellation should be justified to the airliner.

The pilot found himself in a situation with hardly any degrees of freedom, since uncertainty increases complexity and creates interactional problematics and unsettled questions regarding the actor's (here controller's) knowledge and intentions (Mortensen 1997). But, Diagoras Rodios is adding some criteria in his reasoning on the pilot's action, below:

> Considering the conditions and by setting the safety factor above all, the captain acted by rejecting the takeoff, something that most likely I would do, also. But it depends on the exact concern feeling, formed by the incomplete message, and how the personality of the pilot perceives this message.

---

[13] The Airports Council International (ACI) evaluates over 450 airports in 44 European countries (http://www.aci.aero/) and granted the 2014 award, in the "10–25 million passenger" category, to Athens International Airport "Eleftherios Venizelos"(http://www.aia.gr/traveler/). Athens secured this award due to its high economic performance in a very challenging context, its excellent work in redeveloping its traffic base while keeping a strong focus on the quality of service. The top-rated accomplishment sustains its position in the following years.

Diagoras Rodios accepts the likelihood of the same action from his part, but there may be a rebuttal in the safety argument. Not all hesitations are the same; not all pilots perceive signals, indications, and contexts of messages the same. When I probed asking him to explain further the personality of the pilot concept, he said that pilots are like doctors in emergency departments; doctors sort patients for treatment priority and allocation of resources by applying "triage" (Iserson and Moskop 2007; Moskop and Iserson 2007). The sorting of patients for emergency treatment using value-system criteria for human life, efficient use of resources, fairness, and fidelity (among others) provides analogies for the aviation system as a critical environment, especially in emergency conditions. The way that pilots perceive incoming signals demands the abstract thinking, immediacy, and priority handling of triage to select and prioritize the next actions, in Diagoras Rodios' view.

Diagoras Rodios feels that triage, as an occupational skill, should be part of situation awareness response for the pilot. The context determines, to a large extent, the meaning of verbal and nonverbal behaviors. The meaning of a given signal, like the incomplete call of hesitation from the controller, depends on the accompanying behavior with a time factor (before or after). Furthermore, it depends on the interaction which preceded the clearance (and the post-clearance incomplete message), the overall situation awareness for the flight, and the trust developed for the controller or the aerodrome. Trust, in this occasion, starts as the belief that the "other" (the pilot) will take into account what "one" (the controller) is saying (Lin 2001). But how can pilots overcome hesitation or its consequences?

> In the pre-flight flight deck crew briefing there are several textbook cases to abort a take-off. Hesitation is not written in those cases, but there is a bottom line to cover any other case. The sentence is: "apart from above, a take-off can be aborted in any other case considered necessary." The decision is the sole responsibility of the pilot who must declare to "STOP". The first officer, by the time captain chose to act around safety for which he doubted most, ought to help in the procedure. Typically, the relevant training requires, in case there is no complete ATC message, to continue the takeoff procedure.

Diagoras Rodios explains that the flight preparation starts with a preflight briefing where several reasons and causes to postpone a takeoff are tested, for the most part before the inquiry to get clearance and before the clearance is given. Hesitation is not an expected reason as it is not supposed to happen. But, the pilot has the ultimate authority to decide about the airplane and the flight, according to the interpretation of the situation and the surrounding conditions. If so, "stop" can be declared even if the decisive experience from the pilot's point of view is idiosyncratic.

The first officer was compliant in his comments and, since he was also a recipient of the incomplete call, he has the responsibility to assist, if he hears otherwise or speak up in case of a disagreement. Although the letter of the briefing dictates the authority of the pilot, Diagoras Rodios emphasizes the collaborative environment where flight deck cooperation is beneficial and appreciated. The controller is part of the team, Diagoras Rodios, explained in further discussion, so one incident should be kept in a symmetrical not spontaneous reaction (a random event to be discussed, not disciplined) and explained in the controllers' debriefing. Diagoras Rodios redirects the corporate pressure for scheduling and dependability; the takeoff time slot is important to adhere, and, if not, the company wants a written explanation.

So, in their training if no new information is conveyed by the controller after clearance (incomplete information equals to zero information), the takeoff continues (as prompted from socioeconomic parameters of the operational environment and regulations for efficacy); the only reservation is the aviation instinct of the pilot (captain) to overcome the clearance and abort, if he considers so. It is the so-called "buoyancy" in the classic description of Langewiesche[14] (1972). Hesitation added a pragmatic dimension from intended speech to a speech act to abort takeoff, due to the inferential nature of the interpretation of the incomplete message based on background or contextual assumptions and general communicative principles (Grice 1989).

6.4.3 Prudence, Fear, and Situational Awareness: ATC View

The report in the scenario was from the pilot's (captain and first officer) point of view. As hesitation came from the controller and most of the pilots, in further discussion which characterizes the incident as "unacceptable", it would be important to pursuit the evaluation for two different air traffic controllers that were included in Chap. 5 (Sect. 5.4).

The first one is *Ikaria Ikaros*, the most experienced controller who participated in the discussion in Sect. 5.4.1. Ikaria Ikaros comments refer to her experience on hesitation, to her evaluation on the pilots' decision/reaction, and to her insights for air traffic controllers training connecting it with safety[15] at a European level. The following excerpt is addressing those topics:

> The take-off is a very critical phase for the flight safety. I experienced such cases several times in my career, but not harmful ones. The hesitation in this example is rather due to lack of situational awareness of the controller. The pilot did the right thing to safeguard the aircraft. Probably the first officer wouldn't do something else, as there was a feeling of uncertainty and hesitation in the controller's voice, so he said that the pilot was prudent to reject. The hesitation repair depends on the time limits and the evaluation of the danger by the decision maker (i.e. during takeoff if there is something at the end of the runway the pilot may judge more unsafe to reject the take off as time is very limited to react under high speed).

Ikaria Ikaros, unlike other professionals in civil aviation who are categorically denying their own incidents, is open to admit that she has experienced hesitations during her long career, but with no consequences for the flights involved. Instead of

---

[14] *[Buoyancy is] perhaps the most hard-to-get-at skill in the whole art of flying - the sensing of 'lift,' the gauging of the firmness of one's sustentation, the 'feel' a pilot must have for his ship's angle of attack, the ability to know how close the ship is to the stalled condition: This is what pilots used to call the flying instinct* (a Langewiesche description).

[15] It is interesting that the EU Performance Scheme (Com.Reg. N° 691/2010) for the period 2012–2014 introduced reference goals for cost-efficiency, capacity, and environment. Safety was not included in the first reference period but in the second reference period (beyond 2014). Most likely, because Eurocontrol aimed to document the utility and benefits of the policy for transborder ATC and not because it is underestimated. See more at http://ec.europa.eu/transport/modes/air/single_european_sky/ses_2_en.htm (and in Chap. 5).

feeling it as her human failure, she prefers to search her human sense-making at the time of hesitation. She attributes hesitation to the lack of situation awareness from the controller (rather GA type), giving the fact that clearance was already given (and the "Disregard" message confirmed that no anomaly was there). She accepts the pilot's decision since the pilot is responsible for his own aircraft and the copilot did not have any other input but the hesitation cues.

The pilot's decision does not happen in a vacuum, as conditions, time constraints, and speed are decisive factors even when the end of the runway is not exactly clear and the airplane may take off any way. The perceived changes in the tone of voice became part of controllers' updates in training, starting from cases where the controller misinterpreted the change of pitch from a pilot and did not prioritize the flight and accident occurred[16] (Fegyveresi 1997) and going to nonverbal fillers of hesitation identified by Jones (2003) and Howard (2008). But what do controllers learn in training for instructing in critical phases of a flight? Ikaria Ikaros emphasizes the role of the facilitator and the decision-making authority of the pilot:

> During my training as a controller, I have learned how to deal with cancel take-off and how to instruct a pilot that there is an imminent danger for the safe continuation of the procedure, due i.e. the presence of an obstacle in the runway, or a runway incursion. At this very critical moment pilot has the full responsibility to evaluate in seconds the received information and act accordingly, if time permits so. If the aircraft speed is still low, the take-off cancellation can be safely handled but if it is near the critical level or above this, then it is too late to stop the aircraft without serious accident to occur. Pilot has the final decision based on the quick evaluation of received information by ATC. Controller's stabilized voice can be proven helpful, but this is not easy all the time, due to the human nature and the stress occurring in such dangerous situations. As all communications between pilots and controllers are recorded there is an additional fear for not being compliant with the standards phraseologies or the rules, thus imposing more stress (voice uncertainty, low voice, fear, sweating, hand trembling, heart bit etc).

As the discussion with Ikaria Ikaros was repeated several times and I asked her to elaborate further, she stresses the fact that training, for controllers, is focusing on the disruptions that they are aware and oversee when they are about to give clearance for takeoff, i.e., obstacles in the runway like FODs (from Sect. 5.5.3) or a completely unexpected intrusion in the runway, and they do not interfere in flight parameters that they cannot control or are out of their reach and the crew is observing. The airplane in the runway rolling to take off is a critical moment for the flight. The pilot has full responsibility and evaluation time hardly exists, so a message is spoken only if there is substantial content that affects the specific phase in the specific timing.

Parenthetically, as Ifaistos Limnios, one of the pilots mentioned in Chap. 5 (Sect. 5.5) and in Sect. 3.1.1, puts it "I am alone with my soul and my aircraft when in take-off and mostly when I am about to land. I feel the plane as an extension of my muscular system." There are two aspects for the pilot to process the handling of a

---

[16] Refers to the NTSB Accident Report on the Avianca (Airline of Columbia Flight), Boeing 707-321B, HK2016 due to fuel exhaustion, January 15, 1990. Crew language proficiency was identified as a contributing factor in the accident, as priority status was not confirmed. See more at http://www.airdisaster.com/reports/ntsb/AAR91-04.pdf

takeoff cancellation: (i) the technical conditions and flight parameters and (ii) to leave room for decision-making when an unexpected input (if it comes) in case substantial information is transmitted from the controller. The final decision connects the pilot's competency of adjustability, the ability to attune to constraints, with the affordances of aviation with pilot as an agent, merely the conditions of the environment to which the pilot is attuned (Greeno 1994).

That is why the concepts of counting downtime tolerance and instant safety risk evaluation in a change of information are among the skills acquired and maintained with deliberate practice; such practice may lead to the view of pilots (especially fighter pilots who attune to a burst of communication signals and dynamic mission changes) as expert performers, in the way described by Ericsson (2006).

Ikaria Ikaros makes two other important points here; she mentions the impact of a stable controller's voice as a stabilizing factor for the pilot and accents the relationship of hesitation with the fear of recording[17] (Nevile 2006). Flight crews and air traffic controllers operate in a discursive space which is completely mediated, highly regulated, and intense and, as main actors, come from different organizational structures and cultures. For the controller, the captain is a listener in a state of excitement conscious and stimulated from the importance of the moment rolling for takeoff. In the post-clearance phase, any message is an interruption of the ongoing expected sequence of event (till the actual takeoff) and is important both for emotions and for sense-making (Salem 2009). The incomplete hesitation call from the controller was an interruption which generated mistrust in terms of uncertainty. Trust is rebuilt, while interactional exchanges develop in predictable patterns.

So, as Ikaria Ikaros notes, a stable voice with a routine volume intensity, rate, and highness/lowness of your voice is helpful to improve trust; this way the controller "talks the talk" repeating a familiar pattern in oral communication which accentuates an empathic emotion. But, as Ikaria Ikaros admits, controllers are only humans, and human nature has a non-predictable complexity when emotional situations arise from stress and fear. So, it is not realistic to expect ideal controllers' reactions, although training and deliberate practice is always in place.

There are also cultural dimensions in hesitation talk (controller) and hesitation listening (pilot, although sometimes is reversed), as described in the scenario. Two dimensions are more relevant; the concept of tolerance to ambiguity (e.g., What could it mean for the flight if the controller's call is completed while at take off? Should we proceed or not?) for how people are capable to operate even when clarity of what is expected is not there and uncertainty avoidance (e.g., Are we able to decide what to do with the takeoff or do we need to wait for more information?) about how people are able to make up their mind with insufficient information. These are concepts negotiated by Hofstede (2009) and Hofstede and Hofstede (2005).

---

[17] FAA describes the requirement in the "Aeronautical Information Manual/Section 4-1-4: Recording and Monitoring," April 3, 2014 (p. 4–1-1 at http://www.faa.gov/air_traffic/publications/media/AIM_Basic_4-03-14.pdf. ICAO "Procedures for Air Navigation Services/ATM", Doc444, ATM 501, 15th Edn, Nov 22, 2007, has similar provisions.

There is some cultural variability on how aviation professions deal with such concepts as, on average, they follow distinct subcultures. Controllers operate in coordinated groups and need to project the *wider picture* of the air traffic they are managing (Porterfield 1997). Pilots (mostly civil aviation ones), on the other hand, are quite more independent individuals using the airplane in the epicenter of a symbolic world which directs to their assigned destination (Cushing 1994). When the airplane arrives safe, then they continue to the next manifestation of departure-arrival, whereas controllers continue immediately to receive transitional or incoming traffic, in the next moment after a previous clearance.

The second point that Ikaria Ikaros makes has to do with the inherent fear of accountability, due to the recording devices which enable backtracking of their interactions. The regulatory constraints for aviators are a function of the technological configurations and limitations they use and the significance of effective communication for safety (Howard 2008). As flight speech is expected to be scripted from the initial to the closing interaction, in the hesitation scenario, the controller was experiencing a symptom of miscommunication which Ikaria Ikaros associates with a professional accountability fear (due to the growing litigation investigations in the sector), as analyzed by Michaelides-Mateou and Mateou (2010); it is not only that the rushing controller is likely to deviate from some standard phraseology, but it cannot stay in closure since their communication is monitored and recorded.

Controllers know what happened in such hesitation cases and want to use it as an example of empirical informativeness, with selective attention to orality to get information versus the archived recorded data, as analyzed for the use of organizational records by Turner (2010). Air traffic control involves both pilots and controllers on rather equal contribution, for its operation. The workload of both parties requires a filtering process; crew members tend to hear primarily communication that start with the call sign of their aircraft, and controllers tend not to hear and respond to pilot confirmations (readbacks) while issuing clearance/instructions for another airplane. That type of filtering is suggested to happen according to the flight phase, in order to maintain situation awareness and more effective listening.

Ikaria Ikaros mentions that the stressful factors a pilot might experience are valid also for the controller. It connects back to the established safety culture, since air traffic control is in all cases a state-governed organization in which accountability chain, just culture, and blame culture have bureaucratic and organizational challenges to be aligned. Ikaria Ikaros has a suggestion with which she tries to accomplish an institutional response in order to avoid "blame culture" and avoid personifying the situation regarding human error, as shown below.

> Frequent meetings with pilots would create a better understanding of each other's working environment and would help to estimate the workload situations. Although we tried to organize this in our place (Athens International Airport Tower), later on, was not easy to continue due to several reasons. Deviations from standards procedures should be kept at minimum. It is important to remember that human error is in our nature and so unavoidable. Team Resource Management (TRM) is important to better coordinate team during emergency situations.

As workload is a shared reality for both pilots and controllers but not always perceived that way, as they are two different subcultures. The establishment of periodic meetings favors the contact hypothesis (Pettigrew and Tropp 2008). Exposure and positive attitudes are linked together in intercultural communication. When I asked, Ikaria Ikaros, about the reasons they did not continue the meetings, she mainly stated difficulties in scheduling and smooth normality; when no incidents happen, there is no specific motivation for busy individuals to participate in meetings. The safe route is to use the explicit knowledge which comes from the standard operating procedures (training and manuals).

As air traffic control has a socially organized workplace in groups, team resource management (TRM) is an effective way to go forward in further coordination. Ikaria Ikaros stresses that during emergencies (and not only), the optimal of available resources is accomplished at a group and not at an individual level. The air traffic controllers' team is another subsystem with information on floating information, equipment, and people to cooperate.

6.4.4 The Ears and the Eyes of the Pilot on the Ground: ATC View

Putting the same scenario forward, another controller from Sect. 5.4.2, *Oia Santorinis*, describes her case in a realistic, down-to-earth approach. Oia Santorinis perceives her role as informative and trusting for the pilot and points out a clear way to deal with errors. Her quote follows:

> There are several cases where ATC hesitate on the microphone during passing a message to the Captain. That could happen because the ATC may be too stressed with complicated traffic, or even tired and anxious to cope with it. We are trained to keep a calm, strict and precise tone of voice in order to achieve a sense of safety to the aircraft. We are supposed to be "*the ears and eyes of the pilot on the ground*" and our job is to provide them with such information and clearances that are absolutely safe for their cruise. I did not have such experience of a colleague being unsure of what he/she conveys to the pilots. I am of the opinion, and that is something rather common, that even a wrong clearance or instruction should be transferred to the pilot with preciseness, calmness and certainty. Afterwards, the ATC may correct his/her previous clearance without putting any additional pressure to the pilot due to his unsure tone of voice.

Oia Santorinis seems to consider a hesitation mistake not so rare (although she did not have such experience herself or a known colleague), and can be justified from the controller's operating environment, although their training is to insulate them to stay unaffected and calm, with steady regulating voice to inspire a sense of safety. As she mentions, the controllers extend the pilot's senses from the air to the ground, as nerves in an almost neurological bond.[18] A common analogy used from most

[18] The brain, spinal cord, and nerves make up the nervous system. Together they control all the workings of the body. When something goes wrong with a part of your nervous system, you can have trouble moving, speaking, swallowing, breathing, or learning. Someone can also have problems with memory, senses, or mood (more at National Library of Medicine at http://www.nlm.nih.gov/medlineplus/neurologicdiseases.html).

controllers is that the pilot *is the brain in the sky that needs nerves on the ground to land*, and in takeoff the brain is on the ground and the pilot has the nerves till the actual takeoff.

Hesitation on the microphone seemingly connects a mental reservation to speak up after several interruptions of the surrounding interactions from other flights (young controllers feel it as a public speaking drama for them, in the start); it is a case of fragmented attention of the professional-self of the controller who knows that transmitting confidence is a reception of trust for the pilot, similar to the controller's own confidence to automotive tools and channels in his/her work (Lee and Morray 1994). Pressure from air and airport traffic demands repeated interactions and a deluge of data to get responded with sequential information in strict order (each airplane needs to follow a certain sequence of steps, one after the other, in order to arrive to the desired position to move, take off, and land).

The controller must change turns, from one flight to the other, and then come back to the one that is waiting and then to the next and so on (nested loops). It is a case of conversation persistence because the controller cannot stop till resolving or waiting previous calls (Hutchby 2001). Beyond that, complexity arises because there are interdependencies between flights which share the airspace, runway, taxiway resources, and above all the time slots to avoid delays and prioritize routes[19] (like in Saporito 2014). Controllers, often (as much as pilots do), with the multitude of channeled data, signals, stimuli, and reminders are having symptoms of information obesity, as described by Johnson (2012).

Oia Santorinis is arguing on two other very important points; the cooperative role of the controller as an alter ego of the pilot on the ground and the traceable way to repair (and finally correct) errors. She is aware of the air traffic management framework where flights operate and adds that the controller has the role of a safety monitor on the destination point of the flight (and not so much from the departure point of the flight; after takeoff the flight distances itself from the departure point and communication frequency gradually[20]). The controller is on the ground, has more independent variables to configure and control by position, and is expected to relay dependability and safety with the tone of voice. The incomplete message of hesitation is an indication of powerless language which detracts from the controller's ability to show control over the situation, a pattern analyzed by Liska (1992).

Pilots and controllers are users in the computer-supported and technologically mediated collaborative environment of aviation, where communication from one component to the other (and the liveware) generates the flow of information and coordination described by Suchman (2009, 1997; Suchman et al. 1999); it is an

---

[19] That kind of complexity is eloquently described in the *Time Magazine* 2014 article on the "Cancellator," i.e., a flight information system they operate when they need to decide "which flies, where" after repeated cancellations due to harsh weather conditions. The multiplicity of parameters requires the use of "heuristics" as working solutions and not only rule-based criteria. See more at http://time.com/8703/fourth-of-july-flight-cancelllation/.

[20] Of course, the departure point of contact is not underestimated, since several of the parameters and handling are decided and done at takeoff and are decisive about the next phase of the flight. In the case of the ghost plane Helios 522, the manual switch position, the master caution alarm, and the radio frequency set to Larnaca (among other variables) were determinants of the tragedy. The reenactment flight supported the validity of those conclusions from the HAAISB report.

example of the "holon" (the subsystems are wholes and simultaneously parts of other wholes) mentioned in Sect. 6.4.1. Pilots and controllers function at the two ends, on the air and on the ground, and the flight connects them.

Thus, Oia Santorinis incorporates an ideal role implementation for the controller, with the use of a rhetorical conduit metaphor, as a direction of problem-solving technique in flights and meaningful in any context (Lakoff and Johnson 1980); controllers are *the ears and eyes of the pilot on the ground*, and the division of labor for the role in air traffic management demands from them information and clearances as definite and nonnegotiable inputs (Nevile 2004). If the controller voices an erroneous clearance or instruction, the voice has still to be precise and definite, so that it is understood and then corrected with a new message which will be in a low-context and explicit expression.

Oia Santorinis believes that the sequential nature of pilot-controller interaction guarantees consistency when an error is clear and its repair (with a new message) is clear. The controller voices the incorrect message based on the situated understanding of what is happening at the time of the erroneous message with the "local" certainty and rationality (Dekker 2002, 2012).

If the message is precise and calm in tone, it does not add uncertainty for the pilot. The next message which corrects the previous erroneous one is a chunk of a new episode in pilot-controller interaction that is context-specific and not concept-dependent (if looking for the error, blame leads to conceptualization of the situation which caused the error, from the controller's point of view). The approach that Oia Santorinis introduces accounts for the fact that safety is not inherit in the aviation system but people create and implement safety; the pilot will be able to perceive the new message without uncertainty, as the point is not to correct an error with an erroneous nonverbalism like hesitation. This strategy of correcting the error with a clear new message facilitates also the auditing trail (using the recordings), if there is a need to reconstruct the expressed interaction (from an accident investigator's point of view), and adds on to the transparent interactional behavior.

Understanding the pilot, as Oia Santorinis continues, hesitation would create a dependability gap which demands conversational persistence about the "why" of the mistake. Her quote is below:

> If I were a pilot, I would be terrified if I was given the impression by the ATC that he/she is not decisive about the message or the clearance. I would think that something goes wrong and I would be concerned about the safety of the flight. Furthermore, I would be persistent with the ATC to give me more details about the situation and I would ask if there is anything wrong.

Depending on the phase in which the flight and the pilot function, a discussion on causation will not be constructive, since the next step is evolving and the pilot needs guidance to continue and not being engaged in unnecessary discussion. That is why Oia Santorinis suggests the explicit statement to correct the previous message which will restore credibility in the eyes of the crew, as they understand that controllers keep an eye on them to ensure safety navigation in their control area. Oia Santorinis is consistent is her practice (recalling her reply on the first officer's no-talk from Sect. 5.4.2) when telling that the first officer would continue to clarify hesitation in order to assist the captain who is preoccupied with the actual navigation, depending on the takeoff phase and time constraints (see below):

> I believe that there could be a way to make things better after a hesitation message on behalf of the controller. The controller can correct his/her previous uncertain message by him/her being more decisive about the next message. This could reverse the unpleasant feeling of the pilot that the person supposed to "keep an eye on him", the controller, is not Certain for the progress of the flight. The only case where the pilot should ignore the controller is when TCAS system (Traffic Collision Advisory System) is activated, giving instructions to the pilot in order to avoid a collision. In any other case, the first officer should clarify the message conveyed and ask more details from the controller.

Lastly, Oia Santorinis adds a technical enhancement that would justify a deviation from a clearance, as a pilot action. If the traffic collision indicators which operate in the airplane activate the Traffic Collision Advisory System[21] inside the cockpit, then the pilot bypasses any other communication and must apply those instructions to avoid a projected collision. This is an automatic anti-collision system in which inter-transponder communication between aircrafts in proximity is established and provides instructions to avoid a collision and keep the necessary separation distances. The use of those automatic anti-collision systems was an early suggestion in the industry in the 1960s and 1970s and has initiated a discussion about refinements and training after several accidents.[22]

In any case, beyond the actual exchanges of messages, several flight parameters and standards (like the phase of the flight, speed, fuel, weight, weather conditions, etc.) are determinants of the pilot's decision in hesitation and assess if there is even

---

[21] Eurocontrol's skybrary explains in an enhancing safety chapter (boldface added to show the ATC involvement and the importance of the visual channel): *the Airborne Collision Avoidance System II (ACAS II) was introduced in order to reduce the risk of mid-air collisions or near mid-air collisions between aircraft. It serves* ***as a last-resort safety net irrespective of any separation standards.*** *ACAS II is an aircraft system based on Secondary Surveillance Radar (SSR) transponder signals. ACAS II interrogates the Mode C and Mode S transponders of nearby aircraft ('intruders') and from the replies tracks their altitude and range and issues alerts to the pilots, as appropriate. Non-transponding aircraft are not detected. ACAS II works independently of the aircraft navigation, flight management systems, and Air Traffic Control (ATC) ground systems.* ***While assessing threats it does not take into account the ATC clearance, pilot's intentions or autopilot inputs****.Two types of alert can be issued by ACAS II - TA (Traffic Advisory) and RA (Resolution Advisory). The former is intended to assist* ***the pilot in the visual acquisition of the conflicting aircraft*** *and prepare the pilot for a potential RA.If a risk of collision is established by ACAS II, an RA will be generated. Broadly speaking, RAs tell the pilot the range of vertical speed at which the aircraft should be flown to avoid the threat aircraft.* ***The visual indication of these rates is shown on the flight instruments****. It is accompanied by an audible message indicating the intention of the RA. A "Clear of Conflict" message will be generated when the aircraft diverge horizontally.*

[22] Indicative cases are the Japan Airlines near-miss incident (2001); the Überlingen midair collision, between a Boeing 757 and a Tupolev Tu-154 (2002); and the Gol Flight 1907 collision with an Embraer Legacy 600 (2006).

time for intra-crew coordination (for the first officer to talk and confirm) or a clear repair from the controller. For another time, an interactional collaborative accomplishment should lay the ground for an optimal decision-making process by the pilot-in-command, in due time.

6.5 Language Directness, Familiarity Relationship, and Knowledge

Modern aviation operates in an environment organized, constructed, regulated, and smoothly operating through human communicative action. Aviation professionals, particularly crew members and air traffic controllers, have a common understanding of the cooperation and coordination needed to pursuit safety and efficiency in their work and their industry. Aviation communication typically involves a complex interplay between individual cognitive factors and social interactive factors (Cushing 1994). The cognitive mental processes of the speaker or hearer include the interpretive model of the situation, assumed values, judgments, and beliefs, whereas interactive factors include standardized definitions and protocols, cultural requirements, and hierarchical perceptions, among other elements. Both kinds of factors are indispensable when communication messages are verbalized and exchange using the medium of language. Voice has a natural almost biological appeal; it is convenient and, thus, preferred both in human-to-human and human-to-machine communication. Verbal communication uses the vocal signal system with words, both oral and written words. In the current last part of Chap. 6, the narrative moves to verbal communication using language, as the collection of symbols put in words, and words in order (with syntactic rules) to convey meanings in a structure appropriate for the aviation context, its requirements, and particularities.

The way aviation professionals communicate verbally has several attributes of a *private language*, with expressions for what can be known only to the speaker and the close group of similar speakers, unlike the analysis of Wittgenstein (1967). It is private, because of the connection that pilots have with their flight missions and the internalized subculture of the aviator, in which sensation for the aircraft and sensation for being airborne is included. Avgerou and Cornford (1993) used a systems view of language, as an information-carrying mechanism, containing (i) pragmatics as social context, assumptions, expectations,and beliefs that concern all those involved in communication; (ii) semantics as the meaning and knowledge of the signals, words, signs, and their interpretations; (iii) syntactics as the logic and grammar used for the construction and transmission (communication) of words; and (iv) empirics as the codes and the physical characteristics of words in various channels of communication (e.g., pronunciation and proper spelling are included).

In aviation communication, pragmatics and empirics affect situation awareness; pragmatics aligns the intended meaning of the speaker with the perceived meaning of the receiver. As Grice (1989) explains, the receiver is inferring a hypothetical meaning from the speaker and uses the psychological and social context to validate it. This context, in aviation, has very much to do with the situation awareness developing from both sides. In their turn, syntactics and semantics comply with the requirements of control and maximization of efficiency and safety. The order of words in message call must be maintained;

e.g. the standard format to call the air traffic control tower is:
who you're calling, who you are (callsign), [and then] *where you are,* and *what you want.*
Three zero tower is a call *from* the ower *to* the station *three zero* (call sign) and.
the response *tower three zero, go ahead* confirms that the ground station *three zero* is ready to hear the tower.
The order of words is standardized and conducive to "who is calling (to) whom".

This call was the daily first greeting call in the start of each shift when I was the liaison officer between HAF and USN responsible (among other things) for security clearance, takeoff and landing, and freight monitoring for every incoming and outgoing flight in Souda Bay, Crete, Greece. I was in the ground diplomatic station, but the pilot-controller communication had an additional portion of a speech act (confirming a physical act in the cockpit), in the form of a restricted register (Ragan 1997). For example, *turn left to two five zero* contains an instruction (to turn left) and a value (to the direction of two five zero). This is a simple stand-alone unit of information with a parameter and its value in a specific talk turn; if more than one unit are put one after the other, then a complex expression is formed, if needed for a specific sequence of actions in one call. The controller calls a prompting speech act, and the crew confirms (reads back) and performs the act as manifested in the radar screens of both parties.[23]

These units of information qualify as information artifacts (of an oral document status in the end of the flight) with the materiality of their substance in navigating, the institutionalization of their context in aviation talk, and the structure embedded

[23] Air traffic control standard international practice is to monitor airspace using two radar systems: primary and secondary. Primary detects and measures the approximate position of aircraft using reflected radio signals. It does this whether or not the subject wants to be tracked. Secondary radar, which relies on targets being equipped with a transponder, also requests additional information from the aircraft – such as its identity and altitude. All commercial aircraft are equipped with transponders (an abbreviation of "transmitter responder"), which automatically transmit a unique four-digit code when they receive a radio signal sent by radar. The code gives the plane's identity, and radar stations go on to establish speed and direction by monitoring successive transmissions. This flight data is then relayed to air traffic controllers. However, once an aircraft is more than 240 km (150 miles) out to sea, radar coverage fades, and air crew keep in touch with air traffic control and other aircraft using high-frequency radio. A comprehensive description from BBC news at http://www.bbc.com/news/world-asia-pacific-26544554.

in their vocalization (Turner 2010). Moreover, their orality is materialized in the FDR and CVR devices with limited historicity due to the recording loop, unless an accident happens.[24]

An example of the semantic dimension in "airspeak English," or aviation parlance as Cushing (1994) calls it, is the verb *hold* that means to stop what you are doing where as in plain (and colloquial) English,[25] it can also mean continue or maintain what you are now doing.[26] That is why aviation discourse highly regulates symbolic action which must be coordinated, between crew members and air traffic controllers, as the major partners. Aviation discourse has contextual particularities which turns this subset of English as a language for special purposes in Ragan's terms (1994). When the grammatical structure of the sentence matches the speech act with immediacy for a statement, a command, or a question, then language is identified as direct in a natural way.

6.5.1 Directness and Familiarity Scenario

Further investigation of directness in verbal aviation discourse between controllers and crew members (the pilots mostly) is conducted with the employment of another scenario, in order to identify the relation of direct language with cultural or linguistic factors. The scenario is formulated from preliminary discussions for the

[24]Black box flight recorders usually comprise two individual boxes: the cockpit voice recorder (CVR) and the flight data recorder (FDR). Popularly known as "black boxes," these flight recorders are in fact painted orange to help in their recovery following an accident. The CVR would be better named the "cockpit audio recorder" as it provides far more than just the voices of the pilots. In fact, it creates a record of the total audio environment in the cockpit area. This includes crew conversation, radio transmissions, aural alarms, control movements, switch activations, engine noise, and airflow noise. Older CVRs retain the last 30 minutes of an aircraft's flight. **A modern CVR retains the last 2 hours of information**. The newest data records over the oldest data (endless-loop principle).The FDR records flight parameters: pressure altitude, indicated airspeed, magnetic heading, normal acceleration, microphone keying, etc. Microphone keying (the time radio transmissions were made by the crew) is recorded to **correlate FDR data with CVR information**. **The FDR retains the last 25 hours of aircraft operation** and, like the CVR, operates on the endless-loop principle. As FDRs have a longer recording duration than CVRs, they are very useful for investigating incidents and accidents. Material (boldface highlight the recording memory storage) from the Australian Transport Safety Bureau at http://www.atsb.gov.au/publications/2014/black-box-flight-recorders.aspx.

[25]There is a full collection of definitions on the verb in Encyclopedia Britannica's Merriam Webster's dictionary, particularly in the transitive and intransitive form at http://www.merriam-webster.com/dictionary/hold?show=0&t=1409498055

[26]The ambiguity of the verb "hold" contributed to the accident of Air California Flight 336 Boeing 737–293, N468 AC, John Wayne Orange County Airport, Santa Ana, California, February 17, 1981 (NTSB/AAR-81–12). The airplane landed with gear up when attempting to execute a go – around to avoid an airplane on the runway. See more at http://libraryonline.erau.edu/online-full-text/ntsb/aircraft-accident-reports/AAR81-12.pdf.

conduction of the major flight phraseology survey (IATA 2011). It is posted below, in two parts, in the form of a narration from a civil aviation pilot, namely, *Stan Dardman*:

Directness and Familiarity Scenario [*Part a*]
Of course, if I know the ATC person I may bypass the standard phraseology and say "Eh, John this is Tom; what happened, did you hear me? Do you mean it or do I have to wait for descent?" If not, I may just stay to wait in patterns, especially if I do not have a series of corresponding flights in a sequence. I know that safety is the primary concern, but we are also professionals who communicate in discussion terms. I feel it as more natural.

Stan Dardman's narrative introduces familiarity, as a reason for bypassing the standard phraseology, to speak in natural language terms and not in the restricted register vocabulary of air speak English. It seems that he was waiting for a reply in the inquiry of the final approach to land, and since there was a delay to get a reply, he could choose to fly in a holding pattern, if there is no immediate other flight to catch to a new destination. He acknowledges the importance of flight safety but supports the idea of a more tolerant behavior for professionals who know what they are doing even with a nonstandard phraseology that sounds natural. Since Stan Dardman has to work long hours and for years in a mediated and regulated technological environment, human contact was of great value. The question regarding this scenario is asking about the balance between direct and informal language for efficient communication and about the potential authoritative mindset that pilots show over the controllers. The second part of the scenario is below, with the pilot Stan Dardman concluding:

Directness and Familiarity Scenario [*Part b*]
It took me, I could say, more than 1000 flying hours to start feeling safe about myself and specific travel routes (when the weather is in general terms "stable"). I try to include in my social relations people from ATC, as well as other aviation professionals in order to develop a level of familiarity for their personalities.

With the accumulation of expertise in flight hours, travel routes, and self-confidence, as Stan Dardman mentions, one should feel comfortable professionally; i.e. more competent and experienced. While evolving in levels of expertise - in this case the margin was set in 1,000 flight hours- the "expert" may start looking aside of the typical ruling path and realize broader dimensions of the aviation profession. The pilot feels also the need to understand controllers better, starting from a social aspect to facilitate him to conclude on typical personality characteristics of the air traffic controllers as professionals in a community of practice (as explained in Sect. 6.5.2).

A community of practice is a social unit with members (pilots and controllers in this case) who share and learn, based on their common interests in aviation safety and their mutual engagement (often times out of their work schedule). Being a community, it enables the members to collaborate and augment their tacit knowledge and workable practices, as presented in Lesser and Storck (2001).

6.5.2 Phraseology, Professionalism, and Cultural Familiarity

Oia Santorinis, from the young controllers' generation, has also replied in previous scenarios (from Chap.5). She discusses her views on the directness scenario, as well as the relationship-building mindset below.

> I am quite strict about using nonstandard phraseology even if the controller knows the pilot personally. The pilot above implicates that he may have better service by the controller he knows if he speaks to her in an open language. However, this is not the case. Phraseology in our field exists in order to avoid speaking in an open language and avoid misunderstandings. There may be a more familiar tone between two people knowing each other, only when they want to send greetings or call one another with his/her first name to say good morning or hello.

Oia Santorinis, although in her first years in the profession (or even occasionally because of that), is categorically definite about the use of standard phraseology, despite a potential personal connection with the pilot. Being in Greece, a small country with a dense network of airports where the likelihood of repeated flights is high and with not more than three domestic airliners (and a couple of low-cost ones), it is almost inevitable that controllers and pilots gradually acquire familiarity. But, Oia Santorinis is clear on that since it is part of her credibility; she does not accept the implication of alleged preferential service to the pilots she knows (in comparison with those she does not know).

Language is one of the tools controllers use in their professional life and reflects attitudes, commitment, and responsibility. Standard phraseology helps lessen ambiguity of spoken language and facilitates common understanding. Nonstandard phraseology might be simplified, but the meaning may change completely in its intended message, especially when standard keywords like the call sign (e.g., *US 1549*) or numbered locations (turn to *two three zero* or *Aircraft go on to 6000 feet*) are replaced with nonstandard colloquialisms (like *this is me, Tom*) or do not explain what is the reference of specific numbers in the message (which aircraft is mentioned or numbers denoting an altitude, a heading, an airspeed). Nonstandard phraseology, where names are involved, shows familiarity indeed and may shape attitudes (Vanlear 1991).

But that level of familiarity is not compatible with a common level of situation awareness when radio communication is spoken in a shared service line, the party-line communication channel.[27] Since it is not only important what the controller's say but also how it is said, Oia Santorinis admits that familiarity could be expressed with the use of paralanguage with a speech mannerism or tone of voice adjustment without creating unnecessary distractions. A named "hello/goodbye" greeting is a level of professional politeness that she would recognize.[28] She continues to say:

> In any other case, the standard phraseology should be applied putting aside any acquaintances. It is not a matter of authority but safety. Open language does not and should not imply faster service. The controller should be the same to every pilot keeping their professional relation above every personal relation.

Oia Santorinis concludes that standardization is in the essence of the aviation professions and especially in "tower communications" where a restricted register exists. As Ragan (1997) puts it, users of restricted registers must comprehend and produce utterances compliant with the requirements of a very specific situation. After all, open language (in forms of plain language) does not imply faster service but rather more uncertainty and communication noise.

Oia Santorinis moves on to comment on the development of personal (but not inappropriate) relationships with pilots in shifts. Professional awareness of the other's work (the pilot) generates interactional trust and resists to preferential treatment, while professional dependability is enriched with personal familiarity. Oia Santorinis' position is quoted below:

> I agree with the opinion of the above pilot, as I usually do the same thing. For example, when I travel by airplane, I intend to fly inside the cockpit in order to get to know the pilots, their personality, their procedures on board, how they understand the messages of ATCs, the time needed for response to them, their use of language and phraseology. This is very helpful, because if you know the pilots culture could mean that you can interact in a more effective and efficient way. Professional dependability is may suffice but it would be an additional advantage if there was a personal familiarity. To be more specific, by familiarity, I do not mean using open language or chatting on the frequency, but getting to know better pilot culture, education and role in several situations.

---

[27] The "party line" refers to the open radio channel through which all aircraft in a given airspace communicate with air traffic control (ATC), a system that allows pilots to hear both their own clearances and those of the other aircraft. The downside of that is that it increases the workload of pilot functions and multitasking.

[28] It is indicative that in the Malaysian Airlines Flight 370 from Kuala Lumpur to Beijing that disappeared on March 8, 2014, the phrase that was released from the Malaysian authorities was a sign-off "All right, good night" from the cockpit to the controllers; later it was retracted as "Good night Malaysian three seven zero." The formality difference (use of call sign with a greeting and no confirmation) of the two greetings has raised additional questions (e.g., is it a deliberate transmission or not, who is speaking from the cabin, is it a farewell for the flight, etc.) in a case that is still missing, and international investigation was suspended. After a formal request from Malaysia, the Australian Transportation Security Board is leading the search of MH370 in vast areas of the Indian Ocean, although under suspension (More at http://www.atsb.gov.au/mh370.aspx/).

Oia Santorinis offers an insightful point, here, that is very important for the controller to understand the work and the mindset of the pilot, in the realities of their actual "doing the job" (as she accepts pilots would do for the air traffic controllers). She discloses an unconventional small-scale practice of asking to travel inside the cockpit (when she has a travel arrangement) in order to become more familiar with the pilot operations, procedures, time management, and conversational style. When I probed further to her to explain her "intrusion" in the cockpit, she has explained that in domestic flights that last less than an hour (quite common in the Greek FIR), she is just asking that from the crew in the takeoff and landing process. In other flights of Greek airliners abroad, there is a special request from the professional association of Air Traffic Controllers to HCAA, and it is feasible only if (and at the time and for the interval) the cockpit crew finds it acceptable and not disturbing. So, it is not an intrusion but an institutional provision but only when cockpit members allow it (and they are not obliged to).

This way Oia Santorinis becomes familiar with the working culture and training background inside the cockpit and understands better what to adjust on her part. She is a woman in air traffic control and likely to be more expressive and accommodating in conversation,[29] especially when non-routine situations occur, as several of her colleagues observed in their common shifts. This is consistent with the conversational patterns of behavior that are gender-related, as Giles and Street (1985) exemplify in their research. In this manner, Oia Santorinis formulates her personal familiarity, not to shift interactions to open language or start chatting, in order to ameliorate her communication as a controller to the cockpit. Thus, professional dependability coupled with that concept of familiarity generates a collaborative scheme in their work.

This familiarity disposition that Oia Santorinis explains is compatible with the cooperative behavior which (on average) is intrinsic in how the airport tower works with the air traffic controllers (Tavanti et al. 2006). Air traffic control systems have a "socially organized character" (Bannon and Shapiro 1994), and controllers have a strong sense of a community of practice in the larger context of interaction with their environment as human beings (Marti and Scrivani 2003). Oia Santorinis commentary is a profound manifestation of personal responsibility and (collective) work ethics with remarkable self-motivation dimensions.

[29] To add a personal experience in the air traffic control gender discussion, after the first month in my duty in NSA Souda Bay, I have noticed that two Navy women were in night tower operation shifts (from 7:00 pm till 7:00 am in the next day) quite more frequently than that men. I have also noticed that in several emergency landings, the base operations office was asking those two operation controllers to take the microphone. It turned out (after several discussion I had with different members of the NSA command) that the intonation and pitch for the voice of those controllers had a relaxing and mitigating impact in these extreme situation, unlike their male colleagues whose voices did not make any difference.

6.5.3 Minimizing Mistakes, Helpfulness, Personal-Professional Separation

In professional and institutional communication, interlocutors are inclined to follow a collaborative scheme to ensure comprehension, at a certain level of mutuality which is decisive for their work. Oia Santorinis, as stated in Sect. 6.5.1, is actively seeking both the (pilot's) context and the level of understanding needed for her work. Both pilots and controllers communicate with a background of shared knowledge about the English language and the special purposes register of it, about the communication conventions of the pilot-controller confirmation and correction process, and about the operating environment with certain conditions and navigation demands.

In this model of collaboration (Clark and Schaefer 1987), pilot and controllers start with a fiduciary acknowledgment of accepting the information they exchange as appropriate, i.e., as complete, accurate, and timely (but in a real-time evolution). They must both agree that they share the same mental situation model before they continue to the next communication transaction (Billings and Cheaney 1981). That type of harmonization and shared understanding includes persistent confirmation and correction from both sides; i.e., the receiving station (pilot or controller) repeats a received message or an appropriate part thereof (like numbers or directions) back to the transmitting station (controller or pilot) so as to obtain confirmation of correct reception (a readback process). The transmitting station, in turn, repeats the appropriate part of the readback to confirm the correct reception of the readback. So, if A transmits to B, B confirms a readback to A, and A confirms or corrects a hearback to close the communication loop.

In these terms, I went to seek the insights of a fighter pilot with multicultural mission experience to expand on the same scenario of nonstandard phraseology from the pilot *Stan Dardman*. *Eleftherios Venizelos* was the one, with a great deal of experience in NATO reconnaissance missions just after the fighter pilot practice, who had the opportunity to fly over and contact diverse air traffic control towers in 17 European countries (and US bases) which participate in NATO operations. His comments refer to the efficiency of directness and informality and the potential of a power game between controllers and pilots.

The scenario under discussion is the narration of Stan Dardman, about communication with nonstandard phraseology, when the pilot and controller have some type of familiarity (as described in Sect. 7.5.2). Eleftherios Venizelos disagrees with Stan Dardman as shown below:

> I don't agree with the aforementioned comment. When standard phraseology is used then mistakes are minimized. More effective in my experience is when the sender's message achieves the desired purpose. In order to achieve that, you need to give all relevant information, with Standard Terminology, checking for possible ambiguity. If you start using phrases like "do you mean it?" then you are close to a dispute.

Eleftherios Venizelos emphasizes his disagreement with the Stan Dardman's prompt on the use of plain or colloquial English, even though Eleftherios Venizelos comes of a background with strong personal relationships in the pilot subculture. Standard phraseology limits the margin for inferential extensions, for loss of reference, and minimizes mistakes as it is predictable. Effective communication is accomplished when the message of the sender is perceived, and the receiver acts up on as intended to. The restricted window of time applied in aviation demands not only for perception but, also, action and confirmation for the interaction to continue. So, as Eleftherios Venizelos points out, pilot-controller interaction is symmetrical when all relevant information is in place for a specific situation, in standardized manner, so an ambiguity check should become possible.

When explanatory questions are posed, they lead to a possible communication breakdown, in a tit-for-tat interaction between the controller and pilot about which of the two possess the intended meaning. Standardized phraseology demonstrates affiliation with one another in the pilots-controllers group, a phenomenon of linguistic convergence, similar to the adaptations described by Giles et al. (1991). Eleftherios Venizelos does not recognize an issue of authority in the pilot-controller relation, as expressed below:

> It isn't a matter of authority between a pilot and an ATC controller. It is a matter of responsibility. The pilot can request something, and the ATC will work on it, to check if he can grant the request. In my 30-year career as a pilot, the ATC controllers were always willing to help. I can only think an informal discussion with a controller in a small airport, in clear weather conditions with no traffic, in low level. In other words, when all conditions are very nice and under control and there are no safety risks.

Eleftherios Venizelos does not accept a power game, as he emphasizes responsibility. The pilot submits requests for service, and the controller is there to help implement those requests, in terms of availability of resources (air and ground space included or even the frequency capacity permitting, if there is a need of an additional communication channel). A career experience as a fighter pilot in single-seat fighters or in multi-crew reconnaissance aircrafts confirms that controllers are not in other side but in the same side willing to support the mission. Informal discussions are tolerable, in Eleftherios Venizelos' mind, but may happen in ideal flying conditions and in smaller airports/bases, where the window of time is wider and traffic is limited.

As the controller has significant control on the situation on the ground and the scheduling of time slots for incoming and outgoing traffic, a position of power is in place but as a role of service, to do the work. Controllers, as a professional group, and pilots, as their own, come from different subcultures but collaborate to avoid a linguistic intergroup bias, with positive favorable terms for in-group members and negative undesirable behavior of the out-group members (Maas et al. 1989). This accommodating feeling of affiliation is more apparent in smaller aviation communities like the Greek case, in which the density of flight and aerodrome flights, the affinity of aviation with the average layperson, and the international recognition of their professional competence have accomplished professional solidarity atmosphere coupled with a trusting environment (especially when military pilots are in missions or retired military pilots are employed in civil aviation). Eleftherios Venizelos ends his comments about the last part of Stan Dardman's scenario (from Sect. 6.5.1), regarding the social relationships in shifts and the balance between professional dependability and personal familiarity. Eleftherios Venizelos is prone on the separation of professional affiliation from personal relationships, as he states below:

> As a pilot, in multiple Air bases and command centers I neither pursuit social relationships with controllers to increase the feeling of mutual trust nor did conceive an effort from their side to do so. I tend to believe in the personal-professional separation of roles; a key concept was the spirit of cooperation and the friendly atmosphere.

Eleftherios Venizelos says that being in the intense HAF and NATO operating environment, he did not aim to form personal relationships, and he did not have such inquiries from controllers. His mission training and mindset is not directed to create relationships for his everyday life work, as the Air Force has a required level of expertise and dependability for all its members who participate actively in operations. The reciprocal behavior from the controllers to prioritize the professional tasks over personal acquaintance shows a work ethic culture that supersedes organizational patterns, even in the business world where social capital is a key determinant of career in market terms (Krebs 2008). Air Force, especially with the defense-oriented doctrine of HAF, has a mission obligation and a sense of pride which cover all other professional goals.[30] In any case, the trusting environment of collaboration is built on complementarity of roles and constant self-improvement, as shown in the last part below, where Eleftherios Venizelos describes a certain type of guidance which was accommodating to the controllers:

> In my experience as a Tower supervisor in Air Force missions, there have been times when I notice that the controller could not internalize the 3-dimensional projection of air traffic-in Tower without radar for approach. Without interfering to the radio communication, as I was not authorized to do, I have explained that the controller's instruction was not appropriate for the octave of fighter planes which was approaching. The controller did not take into account the diverse speeds of different aircrafts that could generate a cross over if separation margins are not keep proportional. That act has an impact to the projection of routes over time. After this explanation the controller understood how to adjust the given instructions, as mid-air collision was a possibility (especially in those speeds).

In HAF missions, especially when more than a pair of fighters is engaged, military air traffic controllers are accompanied by a high rank officer, a HAF pilot, who

---

[30] Duty-is-my-career mentality was apparent in a collective but anonymous letter (from 35 fighter pilots) which was broadcasted on May 6, 2010, on the occasion of the economic challenges in Greek society (due the economic crisis) and austerity measures and fighter pilot salaries. Their bottom line was "When we are up there flying everything down there is small and we ignore it" (in the top-rated RealFM 97.8 radio, Giorgos Tragkas, "In Athens" radio show). In another occasion, in the National Greek WWII Resistance Day celebration on October 28, 2014, the real-time message "lift the head up" from Captain Stralis the pilot of the HAF Demo Team, during the parade while on flight, stimulated the morale of the people in the Greek society and diasporas (see http://www.protothema.gr/greece/article/421894/sminagos-stralis-eipa-sikoste-to-kefali-psila-ek-merous-tis-polemikis-aeroporias/).

oversees the implementation of the flight or (most often drill). When different types of fighter or attack aircrafts operate together, there may be significant differences in navigation and speed as afforded by each type. Thus, uniform instructions might provoke a risk of separation distance loss when the controller is not able (usually inexperienced) to fully internalize how the screen indicators are pictured in tower screens. Flying missions may collide in real situations, if there are no adjusted instructions for the final approach. Eleftherios Venizelos perceived the mistake while in the tower, but respected the authority of the controller, and explained the situation, and then the controller was able to retract the guiding instructions (in the same mode that *Oia Santorinis* was suggesting in Sect. 5.4.2 for correcting mistakes with responsibility).

The making of the mistake from the controller and the catch from Eleftherios Venizelos, who explained the possibility of erroneous projection of the flight routes, was a constructive confidence-building process that Eleftherios Venizelos includes in the accommodating atmosphere of cooperation in the operating environment of the tower. Eleftherios Venizelos' explanation, using his tacit knowledge in Polanyi's terms (1966), offered the controller a form of information acquisition from a trusted, responsible, and capable pilot supervisor who became the controller's colleague, in the collegial sense of experts (Koskinen et al. 2003).

6.5.4 Home Base, Surrounding Traffic, Intimacy, and Mindfulness

The last response on the same scenario of "social relationships" comes from *Ippokratis Koos*, the helicopter pilot of Hellenic Army Aviation we have met in Sect. 5.5.3. He is more of a team player, despite his leadership position in his battalion, and always tries to inculcate collaboration beyond mere authority as a Colonel of the Army. The last chapter of his career involves the active design of war simulation games for tactical training in NATO missions. In this stage, he must be quite realistic about the actual environment, even beyond the ideal case sometimes presented and used in training manuals.

Ippokratis Koos comments on the scenario of *Stan Dardman*'s bypassing of standard phraseology and professionalism, as well as the tendency to seek for personal relationships. His insights are below:

> The non-standard phraseology communication usually happens at home base and on secondary radio frequencies or low traffic airports, where the environment "seems/sounds familiar". The non-standard communication is true that sometimes facilitate understanding and accelerate a procedure (e.g."*-Hey, Michael, leave the external load north of the tower to get through the ground personnel moving this time on your right. - OK, Ipporkatis, got it* ").

Ippokratis Koos, without any latent thoughts, is clear to distinguish two categories of cases: (i) home base or secondary frequency communication or low traffic airport and (ii) compulsory standardization in the other case and in favor of the surrounding traffic. In the first category mentioned above, Ippokratis Koos is accepting the use of nonstandard phraseology as a fact (and likely prompted) in familiar environments like the home base of a squadron or battalion where the sense of *I know where to go* prevails.

Also, in some occasions the aircraft (it is more common for the helicopters he is commanding, due to flexibility and volatility in their missions) asks for the secondary frequency, in order to avoid a primary frequency congestion and communicate with the controllers (or a mission supervisor) in a more comfortable way (to deal with an emergency or to clarify options and flight choices). The low traffic airport is not a typical case but rather an exception, where air traffic is under control and rather limited interaction is needed. The main reason for the use of nonstandard phraseology is not arbitrary but is functional, since it might facilitate the better understanding of a situation and expedite peripheral supporting activities for specific mission.

In his example, Ippokratis Koos, a certain ground traffic arrangement is agreed with only one interaction with not so explicit use of units of information (*north of the tower*, not numbered coordinates) and rather abstract reference (north of the tower) with an inferential extension (*this time, on your right*). The confirmation *OK, Ippokratis, got it* is in rather plain language but economical and applies to three communicative transaction units *(leave, get through, moving, got it* for all) with respective parameter values that would require step-by-step confirmations (Howard 2008). This instance involves more gut feeling and experience (rather than explicit knowledge from SOPs) learned during observation, practice, and past common experiences and interpersonal contact between *Ippokratis and Michael*, sharing their tacit knowledge (Epstein 1999). It is quite unlikely that the same arrangement would be performed if the two main actors change.

Ippokratis Koos continues to avoid any misinterpretation of his position; there is a level after which the use of nonstandard practice is not sustained and cannot be applied. See below:

> However, the relaxation and the expanded use of it, is dangerous, especially if done between people who do not know each other personally, or under busy air traffic conditions, as there is a risk of misunderstanding, incomplete messages and no confirmations, so that eventually, benefits of compulsory use of standard phraseology are more significant. Furthermore, communication between ATC and a pilot, should be also understood by the surrounding traffic (therefore around Greek airports of heavy traffic, communication is consistently by-the-book and in English language).

Ippokratis Koos stresses that a successful outcome accomplished with a nonstandard language cannot be generalized. The use of nonstandard phraseology cannot be expanded because the actor (pilot or other) feels relaxed about the initial outcome. Tacit knowledge transfer is possible with extensive interpersonal contact, after knowing that "Ippokratis and Michael" know each other (Collins and Hitt 2006). His strong disclaimer is mentioned rather to show the generic case and underline that the exception of nonstandardization is in first category of cases. The risk of misunderstanding, incomplete messages, and omissions of confirmations is a threatening burden of high stake.

Besides that, standardization in language is compulsory because the pilot and the controller are not in a vacuum (or bubble) but in an open airspace environment where the surrounding traffic is approaching, and if they understand the dialogue, they increase their situation awareness. That is why, he mentions, that in Greek international airports which have dense flights, communication is strictly standardized, in ICAO standards.

Ippokratis Koos wants to expand on the case of standard phraseology and restricted register by combining with the last part (part b) of Stan Dardman's scenario, regarding the tendency to build social relationships and balance professional dependability with familiarity. His comments below are mentioned as gap assessment of the first part of his answer with the informal interaction of "Ippokratis and Michael" (italics and explanatory parenthesis added):

> The social relationships between pilots and ATC personnel are something common in the home base environment. The tone of voice, rhythm, and speech pattern between persons that know each other well, leave to pin more things than those described by words. This, definitely, enhances an already good communication as more data transmitted in less time. But, exceeding an intimacy limit in communication, leading to large deviations from standard procedures, should be a red line. For example, it is unacceptable the following communication:
> "- *ES912, after takeoff fly direct to KINETA climbing to 2000'. Incoming traffic, 3 miles east, at 300'.*
> - *OK, guys, thank you.*"
> (Note: No confirmation of instructions and possible risks)

Sociality, for Ippokratis Koos, is inevitable and accepted in the home base environment. The patterns of interaction as well as the paralanguage elements enrich the communication content and information value and materiality of interaction between actors that know each other well (and it happens in the home base). In the same direction, in all international missions, pilots take their aircrafts and maintenance personnel to ensure appropriate operational communication, as if they are in their home base (among other reasons).

This type of language convergence because of intimacy must not lead to broader deviation from standard operating procedures when decisions have to be made (usually in concurrent comparing and contrasting options). Having in mind the informal structure of the interaction between "Ippokratis and Michael", Ippokratis Koos is presenting another view (a projected route after takeoff). The helicopter with call sign *ES912* is instructed to fly direct to a known airport *(KINETA)*, in a certain altitude (*2000*) after takeoff, while there is incoming traffic *(3 miles east, at 300).* The over-abstract response *OK, guys, thank you is* simplistic and out of the standard phraseology, as it has a broad inferential level (where the "OK" goes) and needs confirmation about the instructions and its values per *climbing* step. If one of the instructions is not perceived or misunderstood, the risk increases.[31] The read-

[31] An indicative case for that risk is the fatal accident in Eastern Air Lines 1011, N310EA, Miami, Florida, December 29, 1972. The autopilot was turned off, and the crew did not notice as they had a landing gear issue. The ATC concerned from the loss of altitude asked "ATC: Eastern 401—how are things coming along out there? Captain: Okay – we'd like to turn around now and come back in." as the captain was thinking that the ATC was referring to the landing gear issue they had, just before the plane crashed. More at NTSB/AAR-73-14 report at http://libraryonline.erau.edu/online-full-text/ntsb/aircraft-accident-reports/AAR73-14.pdf

back confirmation is crucial in such type of instructions with more than one value to apply. The harmonization explained in Sect. 6.5.3 is not happening with this interaction.

When noting the contrast that Ippokratis Koos is using to highlight the importance of standardized communication, then the Epstein's (1999) attributes of mindfulness apply. Ippokratis Koos combines explicit knowledge (readily taught and accessible to awareness) with tacit knowledge, learned during observation and deliberate practice (intimacy and repetition of missions), with prior experiences, theories in action (familiarity with the home base environment), and values (social relationship in the home base), and is usually applied more inductively (being able to position equipment and personnel towards the tower and the passage). Apart from that, mindfulness includes a *moment-to-moment self-monitoring* and Ippokratis Koos is conscious of what he is saying as a common rule and what as an exception. Self-monitoring is a control alert and a feedback mechanism to tacit knowledge: Values, peripheral vision, and subsidiary awareness allow for new information and perspectives and trigger curiosity which sustains vigilance.

6.6 Aviation Knowledge Contained in Mother Tongue

The discussion about the voice and language in this book concludes with a short review of issues of language and mother tongue, as reflected, but some of the discussants will be the kick starters for further research. This section organizes the background of language and mother tongue in aviation communication, as collected by side informal discussions with the responders who were involved with the IATA phraseology study (2011), prior to their replies to the specific scenarios of the current study.

Crystal (2003) analyzed how English has become the global language in 1951, primarily because English-speaking countries dominated the flight operations, design, development, and manufacturing of that era. ICAO recommended, as it was not an obligation,[32] English language to be the de facto civil aviation language. Global English became a communication platform for business, science, and technology. In 2008, ICAO standards and recommended procedures (SARPs) required that aviation personnel only in international flights should master a set of word and phrases from the "ICAO Radio Telephony Phraseology".[33] Still there is no prohibition for the use of local (national) languages in domestic airspace.

Furthermore, Crystal recognized conversation registers such as *seaspeak* for maritime use and *airspeak* for air traffic control and aviation. On the other hand,

[32] ICAO Recommendation 5.2.1.1.2 stated that "pending the development and adoption of a more suitable form of speech for universal use in aeronautical radiotelephony communications, the English language should be used as such and should be available on request." This recommendation remains in effect. It is notable that it is not compulsory, as a standard would be.

[33] A quick reference version of that document for commercial air transport pilots can be found at Eurocontrol's skybrary at http://www.skybrary.aero/bookshelf/books/115.pdf

the need for bilingual ATC communications was underlined in empirical cases with aviation professionals. Ragan (1994) characterized "airspeak" English as a controlled language in aviation that is broadly regarded as acultural; in contrast cultural factors and conversation interaction do influence specialized communication.

Airspeak wording is a controlled language with a singular context focusing on human-machine interaction and air transport safety conduct. But most aviation accidents are attributed to human error, with the large majority of it being communication problems (Billings and Cheaney 1981). Problems of communication have been found to be a contributing or (sometimes) causal factor for many critical incidents as well as near and actual accidents in aviation (Cushing 1994; Helmreich et al. 1999; Jones 2003; Krifka et al. 2003; Silberstein and Dietrich 2003; Tajima 2004). In terms of language use, Cushing (1994, 1995) analyzed problems of reference, repetition, inference, ambiguity, and homophony in the English language. Jones (2003) spoke about *symptoms of miscommunication* (as violations of procedures and instructions) and communication-specific problems for language interpretation and phrasing.

On the other side of language use, Jones puts phraseology defects and implausible word choices. As a result, several categories of problematic communication were identified by Howard (2008): incorrect information, failure to provide information, request for repetition, and correction to repair a previous communication error. These categories are manifested in the use of English language in aviation communication and are accompanied by linguistic elements in the use of language: the context and expectations (what to hear instead of what is actually said), speech intelligibility (based on the phonological features of a mother tongue), paralinguistic factors (analyzed in the start of the chapter), aviation jargon (leading to ambiguity), word resemblance (homophones, homographs, homonyms, and synonyms), and code switching (changing languages in the course of communication).

6.6.1 Mother Tongue Scenario and Code Switching

Code switching is the element discussed in this section, as the alternation between languages or dialects of the same language. Code switching occurs even in English between plain English and technical jargons (which creates subsets of English for special purposes). One of the goals in the study of this book is to investigate the role of code switching from English to Greek and its causes and impact as perceived by the real knowledgeable users of the languages (pilots and controllers). The relevant scenario was developed in a briefing discussion in HCAA Flight School, from a civil aviation pilot, namely, *Palios Ellinikos*, and is the following:

Mother Tongue Scenario
ICAO English may be the spontaneous choice when I follow habitual in-flight processes. (without too much processing in my mind). I know there is a difference in FAA when I fly in the US, as recommended phraseology is not the same. Sometimes, I feel as I am a "recorded voice" in action when I apply the routine SOPs. In contrast, when I have to do something more that the standard checklists, I find myself using Greek language and mostly Greek words inside English structures.

Palios Ellinikos is describing, here, a code-switching process in multiple directions; ICAO English have several differences with FAA English (as shown in Appendix D of this book), English come as a default language, routine processes are close to automated passive functions, and checklist marking is like a dialogue mix of Greek and English. Taking this scenario as a reference, the question posed concerns the role of the Greek mother tongue "when communicating" and the comparison of its use in emergencies rather than in routine flights.

6.6.2 Diversity Responding to the Mother Tongue Scenario

Let's see several responses collected from the discussants who participated in this study. I encouraged each responder to recall flights when mother tongue was used. The probing questions were about *what is accomplished when communicating with the mother tongue rather than with standards phraseology.* Also, is it different in emergencies in comparison with routine flights, with the experience of the world around the flight (a GA type of awareness), as well as of the microenvironment of the cockpit, and the need to report and act? *Ikaria Ikaros*, the most experienced controller in the discussion, prompts the following response:

> Yes, definitely, during emergencies, mother language might be proven very helpful as you do not have to think about structure, right words, it is spontaneous. Communication in English is not always easy, especially when you must speak in open language, you might not know all you want to say in the best way.
> I am trying to speak always in English during my work, but it happened to me to choose mother Language to overcome an emergency, when pilots are Greeks. I might need more training in English for such cases, but I wonder if I would ever be able to speak a foreign language as I speak my mother tongue, during emergency cases, when stress prevails.

She emphasizes that although her motto is to speak English to all, when there is an emergency and more information are needed, she prefers the Greek mother tongue with Greek pilots.

Oia *Santorinis*, the enthusiastic and language-sensitive controller, recalls an example where code switching, more than once was helpful. See below:

> On the other hand, when situations out of the ordinary occur, there is standard phraseology that should be applied but does not cover many aspects of the problem. As a result, it is inevitable to deviate from SOPs and use open language showing flexibility and not trouble the pilot who is not able to remember phraseology but feels more comfortable to speak in an open language giving more details and information about the condition either of the plane or the flight. When I was working in Thira, a VFR flight was approaching the airport and the pilot was French. His English was poor and while he was approaching he declared that he had lost his way. In that case, I started talking to him in an open language in order to make him feel more safe and comfortable, using simple English words. He sounded from his voice that he was confused and even scared so when I saw him, approximately 8 miles from the airport, I used one or two phrases in French in order to show him that I was able to speak to his language with which he was the knowledgeable with. It seemed to be helpful because afterwards he sounded more comfortable and confident. Finally, he landed safely.

Oia Santorinis notes that SOPs seem not to cover all eventualities and that is why they are always in a dynamic upgrade when incidents happen. She recalls a flight approaching with a French pilot who sounded confused; she changed to open English to facilitate him and then to French in the final approach so that he would feel more comfortable, safe, and not confused, and it worked.

Eleftherios Venizelos, the fighter pilot with multinational experience, adds on this scenario:

> Bypassing the standard phraseology and code switching in Greek happens when I must request for a combination of things, and it would be easier for both parties to explain it in Greek (with several English jargon in and out). In HAF it is likely to happen to save time from repetitions, and clarifications. A typical example in HAF is that weather permitting, fighter pilots need to save time and fuel for more drills instead of more speaking. Of course, in emergencies it is more efficient to address in the mother tongue. In dog fights there is a mixture of several verbs in imperative and English jargon in between.

Eleftherios Venizelos states that code switching maybe time-consuming when a complex message is transmitted and the way to avoid repeated questions and clarifications. For HAF pilots in drills, to save time is crucial, and the mother tongue applies better (especially in national airspace where threat exists). Mother tongue is also the choice when in an emergency, and code switching in a hybrid mode happens in dogfights.

Lastly, *Ippokratis Koos*, the helicopter pilot of Hellenic Army, brings several important comments in the discussion although he is using the mother tongue more. See below:

> The above text perfectly [in the original scenario] describes what I feel about the use of English and Greek mother tongue. In flights within Greece, routine communications permit the "luxury" of formality using English language. However, in stressful situations (like emergency procedures), when what is urgent is the immediate transmission of a clear message, and the full understanding of it by the receiver, all employ the native language. On flights abroad, or when there is a strong traffic around, the use of English is one way, usually burdening the pilot's workload, depending on his ability to use the English phraseology.

He feels that Palios Ellinikos' scenario is exactly what his own experience in helicopters is. Routine flights have the advantageous option to use English (as for him it shows ideal conditions), but when a stressful situation is evolving, and understanding is crucial for both sides, the mother tongue prevails. In international flights, standard English is used but always as a cumulative factor in pilot's workload and situation awareness.

The common denominator of these replies (and others reviewed) is that mother tongue affects the way pilots and controllers perceive messages and situations, and consequently is a determining factor of their situation awareness (as shown in Sect. 4.3.2 and the example of multinational force of NATO). In emergencies, their mother tongue, the Greek oral language, prevails, but more research is needed to connect emotions with language use and comprehension in extraordinary situations (Valeontis 1999). Lastly, code switching seems to require more exploration, in diverse conditions, and language and mother tongue seem to hold an important part of the programming of the mind (Katerinakis et al. 2013).

6.7 A Knowledgeable Review of Voice

Chapter 6 is the transition from the phenomenon of silence (in Chap. 5) to the use of nonverbal communication, hesitation encounter, and verbal communication in aviation. The chapter flows from silence to paralanguage and verbal communication. Analysis starts with the landmark case of the Helios 522 ghost plane, in which the scrambled F-16 pilot explains the Renegade response, in an effort to communicate with a silent aircraft, as well as his insights and agony. The fighter aircraft, itself with all its technical and physical affordances, was used as a communication tool.

Then the hesitation scenario (as introduced in Sect. 3.2.2) is used with an incomplete tower call that was in a hesitant voice, after a takeoff clearance was received. This scenario is the main inquiry under the question *what would you have done and why* for pilots and air traffic controllers who offer their own reading and evaluation. This question is commonly used in Callback reports in NASA's Aviation Safety Report System, and the reader had the chance to virtually interact with the responders through the flow of the text with scenarios and experts' comments.

Fighter pilots argue for the primacy of safety in comparison with any other criteria that may apply. Controllers' messages should be immediate and clear to avoid uncertainty and misinterpretation. Controllers should operate like the "eyes and the ears" in the ground when the pilot is in-flight airborne. Hesitation might be a result of fear and incomplete situation awareness. The pilot's interpretation depends on a multitude of factors which assemble to "triage," to the so-called aviation instinct, and to the Langewiesche's "buoyancy." Human error may occur in a bad moment and should be accepted in order to get it corrected.

Then the chapter introduces the scenario of directness in language, informality, and social relationship building when exercising communication between pilots and air traffic controllers (in Sect. 6.5.1). This scenario continues with the addition of the importance of understanding the "others" (pilots for the controllers and controllers for the pilots) and their personality traits.

Professional ethics and self-monitoring as defined by Epstein (1999) cultivate the expertise set of tacit knowledge (Doak and Assimakopoulos 2010). The importance of standardized communication is part of a social and collaborative operating environment with the air traffic controllers.

In the last part of the chapter, after an overview of language and English use in aviation communication, an additional scenario is introduced to lay the foundation for future research in pilot-controller communication; the mother tongue as a primary code-switching element (Sect. 6.6.1). The modern aviation system requires the maximum amount of information expressed with the minimum of effort and consuming the minimum of time; so, in several aspects it has been "McDonaldized" in terms of Ritzer (2013).[34]

The distinction between routine and emergency situations is foundational for the mother tongue selection in all cases. The persistence in the use of mother tongue has traces of the linguistic relativism notion (Whorf 1956). Mother tongue seems to provide the evidentiality about the reported facts. Airspeak Greek, analogically to Airspeak English, could construct a concept of language security and trustworthiness for national public policy in the busy Greek airspace.[35]

References

Argyle, M. (1988). *Bodily communication* (2nd ed.). London: Routledge.

Avgerou, C., & Cornford, T. (Eds.). (1993). *Developing information systems: Concepts, issues and practice*. London: McMillan Information Systems Series.

Bannon, L. & Shapiro, D. (1994). Opening up ATC work: Behavioural cognitive and sociological perspectives in Proceedings of the 21st Western European Association for Aviation Psychology (WEAAP), Dublin.

Beier, E. G. (1974). Nonverbal communication: How we send emotional messages. *Psychology Today, 8*, 52–59.

Billings, C. E./Cheaney, E. S. (1981). Information transfer problems in the aviation system. (NASA TP 1875) Moffet Field: NASA- Ames Research Center.

Burns, K. L., & Beier, E. G. (1973). Significance of vocal and visual channels in the decoding of emotional meaning. *Journal of Communication, 23*(March), 118–130.

[34] McDonaldization refers to the reconceptualization of rationalization and scientific management, in the foundation of homogenization with efficiency, calculability, predictability, and control. Ritzer uses the fast-food restaurant as rather representative contemporary paradigm in organizing processes. Like the Henry Ford's pioneering concept of assembly lines in 1913, taking a task and breaking it down into smaller tasks. This is repeated until all tasks have been broken down to the smallest possible level for an efficient operation. Irrationality and other side effects of over-rationalized systems are part of criticism for Ritzer's approach.

[35] Greek aviation and related phenomena consist a seminal case, in a spatial configuration with the density of more than 40 airports in a population of less than eleven million people, with thousand flights (in the busy period) per day and a high-profile Air Force. It is noteworthy that in the latest multinational air defense drill ("Blue Flag" 2017 at the Uvda Air Base, Israel) for composite Air Operations (COMAO), with fighter planes from Israel, Greece, France, Germany, Italy, Poland, India, and the USA, the Greek F-16 formations were selected to be in command of daylight and nightly missions.

Chute, R. D., & Wiener, E. L. (1996). Cockpit/cabin communication II: Shall we tell the pilots? *The International Journal of Aviation Psychology, 6*(3), 211–231.

Clark, H. H., & Schaefer, E. F. (1987). Collaborating on contributions to conversation. *Language and Cognitive Processes, 2*, 19–41. Reprinted in R. Dietrich & C. F. Graumann (Eds.), *Language processing in a social context*. Amsterdam: North Holland, 1989.

Collins, J. D., & Hitt, M. A. (2006). Leveraging tacit knowledge in alliances: The importance of using relational capabilities to build and leverage relational capital. *Journal of Engineering and Technology Management, 23*(3), 147–167.

Crystal, D. (2003). *English as a global language* (2nd ed. First ed., 1997). Cambridge: Cambridge University Press.

Cushing, S. (1994). *Fatal words: Communication clashes and aircraft crashes*. Chicago: The University of Chicago Press.

Cushing, S. (1995). Pilot-air traffic control communications: It's not (only) what you say, it's how you say it. *Flight Safety Digest, 14*(7), 1–10.

Dekker, S. W. A. (2002). Reconstructing human contributions to accidents: The new view on error and performance. *Journal of Safety Research, 33*, 371–385.

Dekker, S. W. A. (2012). *Just culture: Balancing safety and accountability* (2nd ed.). Farnham: Ashgate.

DeVito, J. A. (2011). *Human communication the basic course* (12th ed.). Boston: Allyn and Bacon.

Doak, S., & Assimakopoulos, D. (2010). Tacit knowledge: A needed addition to SOPs, in a forensic science environment. *Forensic Science Policy & Management: An International Journal, 1*(4), 171–177.

Dominguez, C., Vidulich, M., Vogel, E., & McMillan, G. (1994). *Situation awareness: Papers and annotated bibliography*. Armstrong Laboratory, Human System Center, ref. AL/CF-TR-1994-0085.

Endsley, M. R. (2004). Situation awareness: Progress and directions. In S. Banbury & S. Tremblay (Eds.), *A cognitive approach to situation awareness: Theory, measurement and application* (pp. 317–341). Aldershot: Ashgate.

Epstein, R. M. (1999). Mindful practice. *JAMA, 282*(9), 833–839.

Ericsson, K. A. (2006). The influence of experience and deliberate practice on the development of superior expert performance. In K. A. Ericsson, N. Charness, P. Feltovich, & R. R. Hoffman (Eds.), *Cambridge handbook of expertise and expert performance* (pp. 685–706). Cambridge: Cambridge University Press.

Fegyveresi, A. E. (1997, April 27–May 1). Vocal cues and pilot / ATC Communications. In R. S. Jensen & L. A. Rakovan (Eds.), *Proceedings of the Ninth International Symposium on Aviation Psychology* (Vol. 1, pp. 81–84).

Gibson, J. J. (1979). *The ecological approach to visual perception*. New York: Houghton Mifflin.

Giles, H., & Street, R. L. J. (1985). Communicator characteristics and behavior. In M. L. Knapp & G. R. Miller (Eds.), *Handbook of interpersonal communication* (pp. 205–261). Beverly Hills: Sage.

Giles, H., Coupland, N., & Coupland, J. (Eds.). (1991). *Contexts of accommodation: Developments in applied sociolinguistics*. Cambridge: Cambridge University Press.

Greeno, J. G. (1994). Gibson's affordances. *Psychological Review, 101*(2), 336–342.

Grice, H. P. (1989). *Studies in the way of words*. Cambridge, MA: Harvard University Press.

Hall, E. T. (1963). A system for the notation of proxemic behavior. *American Anthropologist, 65*, 1003–10026.

Hall, E. T. (1966/1990). *The hidden dimension*. New York: Anchor Books (Reprint).

Hellenic Air Accident Investigation and Security Board (HAAISB). (2006). Helios Airways Flight HCY522, Boeing 737-31S at Grammatiko Hellas on 14 August 2005. Athens: Ministry of Transportation & Communications.

Helmreich, R. L., Merritt, A. C., & Wilhelm, J. A. (1999). The evolution of crew resource management training in commercial aviation. *International Journal of Aviation Psychology, 9*(1), 19–32.

Hofstede, G. J. (2009). Research on cultures: How to use it in training? *European Journal of Cross-Cultural Competence and Management, 1*(1), 14–21.

Hofstede, G., & Hofstede, G. J. (2005). *Culture and organizations: Software of the mind* (2nd ed.). New York: McGraw-Hill.
Howard, J. W., III. (2008). 'Tower, am I cleared to land?': Problematic communication in aviation discourse. *Human Communication Research, 34*(3), 370–391.
Hutchby, I. (2001). *Conversation and technology: From the telephone to the internet*. Oxford: Polity Press- Blackwell Publishers Ltd.
International Air Transport Association (IATA). (2011). *Pilot and Air Traffic Controllers Phraseology Study (IATA- IFALPA- IFATCA)*. Canada: IATA Reports and Publications.
Iserson, K. V., & Moskop, J. C. (2007). Triage in medicine, part I: Concept, history, and types. *Annals of Emergency Medicine, 49*(3), 275–280.
Johnson, C. (2012). *The information diet: A case for conscious consumption*. Sebastopol: O'Reilly Media.
Jones, R. K. (2003). Miscommunication between pilots and air traffic control. *Language Problems and Language Planning, 27*(3), 233–248.
Katerinakis, T., Papadopoulos, N., & Zouroudi, D. (2013). Aviation terminology: Literal sense, safety and semantic-oriented translatology in Greek aviation communication. In P. Kontos, A. Anastasiadi-Symeonidi, M. Katsogiannou, & K. Valeontis (Eds.), *Proceedings of Hellenic language and terminology, Athens, Greece: Hellenic Society for Terminology Publications*.
Knapp, M. L., & Judith, A. H. (2010). *Nonverbal communication in human interaction* (7th ed.). Boston: Wadsworth Cengage.
Koskinen, K. U., Pihlanto, P., & Vanharanta, H. (2003). Tacit knowledge acquisition and sharing in a project work context. *International Journal of Project Management, 21*(4), 281–290.
Krebs, V. (2008). Social capital: The key to success in the 21st century organization. *International Association for Human Resources journal, 12*(5), 38–42.
Krifka, M., Martens, S., & Schwarz, F. (2003). Group interaction in the cockpit: Some linguistic factors. In R. Dietrich (Ed.), *Communication in high risk environments* (pp. 75–101). Hamburg: Helmut Buske Verlag.
Lakoff, G., & Johnson, M. (1980). *Metaphors we live by*. Chicago: The University of Chicago Press.
Langewiesche, W. (1944/1972). *Stick and rudder: An explanation of the art of flying*. New York: McGraw-Hill.
Lee, J. P., & Moray, N. (1994). Trust, self-confidence and operators' adaptation to automation. *International Journal of Human-Computer Studies., 40*, 153–184.
Lee, J. D., & See, K. A. (2004). Trust in automation: Designing for appropriate reliance. *Human Factors, 46*(1, Spring), 50–80.
Lesser, E. L., & Storck, J. (2001). Communities of practice and organizational performance. *The IBM Systems Journal, 40*(4), 831–841.
Lin, N. (2001). *Social capital: A theory of social structure and action*. Cambridge: Cambridge University Press.
Liska, J. (1992). Dominance-seeking language strategies: Please eat the floor, dogbreath, or I'll rip your lungs out, O.K.? In S. A. Deetz (Ed.), *Communication yearbook* (pp. 427–256). New Brunswick: International Communication Association.
Maas, A., Salvi, D., Acuri, L., & Semin, G. R. (1989). Language use in intergroup contexts: The linguistic intergroup bias. *Journal of Personality and Social Psychology, 57*, 981–993.
Marti, P., & Scrivani, P. (2003). The representation of context in the simulation of complex systems. *Cognitive Technology, 8*(1), 32–42.
McIntosh, D. N. (1996). Facial feedback hypotheses: Evidence, implications, and directions. *Motivation and Emotion, 20*(2), 121–147.
Michaelides-Mateou, S., & Mateou, A. (2010). *Flying in the face of criminalization; the safety implications of prosecuting aviation professionals for accidents*. London: Ashgate.
Moore, N., Hickson, M., III, & Stacks, D. W. (2010). *Nonverbal communication: Studies and applications*. New York: Oxford University Press.
Mortensen, D. C. (1997). *Miscommunication*. Thousand Oaks: Sage Publications.

Moskop, J. C., & Iserson, K. V. (2007). Triage in medicine, part II: Underlying values and principles. *Annals of Emergency Medicine, 49*(3), 282–287.

Nevile, M. (2004). *Beyond the black box: Talk in interaction in the airline cockpit*. Aldershot: Ashgate.

Nevile, M. (2006). *Communication in context: A conversational analysis tool for examining recorded voice data in investigations of aviation occurrences*. Report prepared for the Australian Transport Safety Bureau. ATSB Research and Analysis Report B2005/0118.

Pettigrew, T. F., & Tropp, L. R. (2008). How does intergroup contact reduce prejudice? Meta-analytic tests of three mediators. *European Journal of Social Psychology, 38*(6), 922–934.

Pittenger, R., Hockett, C. F., & Danehy, J. J. (1960). *The first five minutes: A sample of microscopic interview analysis*. Ithaca: Paul Martineau.

Polanyi, M. (1966). *The tacit dimension*. London: Routledge & Kegan Paul.

Porterfield, D. H. (1997). Evaluating controller communication time as a measure of workload. *International Journal of Aviation Psychology, 7*, 171–182.

Ragan, P. H. (1994). Subject specialism and general English in aviation ESP. *TESOL Matters, 4*(4), 7.

Ragan, P. H. (1997). Aviation English: An introduction. *The Journal of Aviation/Aerospace Education & Research, 7*, 24–36.

Ritzer, G. (2013). *The McDonaldization of society: 20th anniversary edition*. Los Angeles: Sage.

Salem, J. P. (2009). *The complexity of human communication*. Cresskill: Hampton Press.

Saporito, B. (2014, February 20). Airport confidential: Inside the strange world of airline cancellations. *Time Magazine*, at http://time.com/8703/fourth-of-july-flight-cancelllation/.

Silberstein, D., & Dietrich, R. (2003). Cockpit communication under high cognitive workload. In Dietrich (Ed.), *Communication in high risk environments, Special issue 12 of Linguistische Berichte* (pp. 9–56). Hamburg: Buske.

Suchman, L. (1997). Centers of coordination. A case and some themes. In L. Resnik, L. Saljo, C. Pontecorvo, & B. Burge (Eds.), *Discourse tools and reasoning. Essays on situated cognition*. Berlin: Springer.

Suchman, L. (2009). *Human- machine reconfigurations, plans and situated actions*. Cambridge: Cambridge University Press.

Suchman, L., Blomberg, J., Orr, J. E., & Trigg, R. (1999). Reconstructing technology as social practice. *American Behavioral Scientist, 43*, 392–408.

Tajima, A. (2004). Fatal communication: English in aviation safety. *World Englishes, 23*(3), 451–470.

Tavanti, M., Marti, P., & Bourgois, M. (2006). Tacit knowledge and frugal artifacts: A challenge for technology mediated collaboration. In *ECCE '06 Proceedings of the 13th European conference on cognitive ergonomics: Trust and control in complex socio-technical systems* (pp. 105–108).

Trager, G. L. (1958). Paralanguage: A first approximation. *Studies in Linguistics, 13*, 1–12.

Tsolakis, A. D. (2013). *Dare! A way of life (Tolma! Enas Tropos Zois)*. Athens: 11Aviation Publications.

Turner, D. (2010). Orally-based information. *Journal of Documentation, 66*(3), 370–383.

Valeontis, K. (1999). *The abandoned spoken Greek language* (I Egkatalimmeni Proforiki Elliniki Glossa). Hellenic Society for Terminology/ELETO-PEEL. Athens: ELETO Eectronic Publications.

Vanlear, C. A. (1991). Testing a cyclical model of communicative openness in relationship development: Two longitudinal studies. *Communication Monographs, 58*(1991), 337–361.

Whorf, B. L. (1956). The relation of habitual thought and behavior to language. In J. B. Carroll (Ed.), *Language, thought and reality: Selected writings of Benjamin Lee Whorf* (pp. 134–159). New York: Technology Press of M.I.T. and Wiley.

Windahl, S., & Signitzer, B., (with Olson, J. T.). (1992). *Using communication theory: An introduction to planned communication*. Thousand Oaks: Sage.

Wittgenstein, L. (1967). *Philosophical investigations*. Oxford: Basil Blackwell.

Chapter 7
Flights Go On, Inquires Pass Through

7.1 Intended Advancements of the Knowledge Base

A successful flight is a systemic result of the interactional accomplishment of human performance. This book is following a communication orientation via the basic model of communication (Shannon and Weaver 1949) and notions from the helical model of communication (Dance 1967). Every interaction is based on a first event, time is important, and communication process is a product of what actors have learned, inward, in ways that permit skill and event growth and discovery of knowledge. The nature of flight communication, which considers meaningful information as the specific selection of a signal from the noise, extends the basic model towards the interhuman communication network (Leydesdorff 2016). In this sense, cockpit speech acts describe the special character of cockpit conversations with internal (on board) participants and the external ones, most notably the air traffic controllers. All participants are considered actors who follow the basic model of communication. Additional supporting literature on speech acts, actor-network theory, trust and trust in automation, as well as rules and roles and identity in the workplace informs the discussion and information processing in the current book. Three major streams of literature are expected to advance with communication-oriented contributions and the resultant (social) knowledge dimensions of this book:

- Systems analysis on coordinating entities and holons (Checkland 2006; Zexian and Xuhui 2010; Ostrom 2005; Cushing 1988; Leveson 2012; Gasson 1999, 2005). Holons are structures composed of subassemblies, in layers of part-whole units of complex adaptive systems. A whole system at one level is a part of a system at another level. In the soft systems approach, the role of the human user is highlighted.
- Human factors analysis of flights as decision-making processes, where human error is considered a system property and safety, is the outcome of coordination

T. Katerinakis, *The Social Construction of Knowledge in Mission-Critical Environments*, Innovation, Technology, and Knowledge Management,
https://doi.org/10.1007/978-3-319-91014-7_7

of individuals (Loukopoulos et al. 2001, 2003, 2009; Barshi 1998; Dismukes et al. 2007; Dekker 2002, 2010, 2011, 2012; Edwards 1988; Endsley 1988; Ericsson 2002).

- Culture in organizational operations, behavior and decision-making with aspects of authority and the parameters of safety, just culture, and the relativity of mother tongue in perception (Merrin 1996; Helmreich and Davies 2004; Helmreich et al. 1999; Hofstede and Hofstede 2005; Hayward 1997; Wenger et al. 2002, 2009; Wegner 2002).

Contributions to systems theory and human factors literature are highlighted in the following sections (with the framework of representation of Chaps. 2 and 4). Cultural components, more closely connected with communication, are pointing to a discussion for further research implications closer to empirical evidence.

The modern aviation system operates as a communication process constructed, organized, and regulated through human actions. Aviation safety is and will remain one of the central concerns of our times for all people living in this planet. Empirical data demonstrate that communication problems have indeed cost lives or were crucial even when no fatalities occur. Comprehensive studies of the role of communication in aviation incidents are limited, but they demonstrate how incident data provide insights to accident causes. In this study, human communication takes the lead in occurrences, incidents, and accidents in flight situations.

Thus, a systems approach that is concerned with information flow and sender-receiver, feedback, and noise perception in the cockpit environment is a contributing factor in the current analysis. Human factors analysis of cognitive interactions in mission critical environments offers another ground to draw on how participants implement their interaction in time-constrained conditions. Aviation human factors research describes the goal of all pilots *to get people from point A to point B, without disturbing or killing them.*

Discourse analysis literature provides a framework of understanding interaction, turn-taking, and implementation of processes via communication. Interaction phenomena in the case of cockpit conversation extend common encounters of turn-taking and routine discussions to pressing operating conditions. In these conditions, human participants need to have established interaction relationships and apply disciplined actions to correct errors and follow consequential steps and accomplish cockpit identities and roles.

Communication Implications Human communication is crucial in order to accomplish a (safe) flight mission. The value-added contribution has been to infuse communicative components to, both, human factors and systems theoretic approaches to aviation, in a way that has not been done before by communication scholars.

There are at least four of those components that document the accomplished theoretical contribution: (i) the deconstruction of silence phenomenon in multiple dimensions, as part of an interaction and accountability process; (ii) the synthetic proposition of a voice-categorical label which consolidates paralanguage and hesitation and nonverbal and verbal attributes, in a communication channel applicable in aviation (the complementary character (for the verbal or nonverbal message) of hesitations connects with different vocalic realizations: filling pauses, word lengthening, repetitions, and speech repairs); (iii) the incorporation of situation awareness, with its local, transitory, and global views, as a component to transform communication to knowledge models; and (iv) the revisit of mother tongue as a non-conflictual but complementary communication tool which may facilitate linguistic security, as language carries knowledge, instead of competing with the lingua franca in aviation, the topical standardized English.

It is important to have in mind that aviation industry pays attention to each sole flight instance, since an incident, event, recommendation or warning observed in a single flight could modify substantial practices and technologies in the sector. Flight operating manuals describe an ideal flight situation, with information presented as sequential, whereas the real-flight operating environment is quite more interactive and pushes communication in the foreground. Overlapping of tasks, interrupting of tasks and communication, and taking the turn to complete each activity are happening in any random flight. Cockpit crew has to interact with a range of human actors on the ground and in the air. These actors send critical information (and hearbacks) to the crew, expect information (and readbacks) from the crew, and affect the timing, sequence, and performance of the crew's tasks with their demands.

A flight is now represented as an act of composite conversation that combines culture with communication of co-present, mediated, and cooperative actors. Cockpit communication, in this study, extends the physical context of the flight deck with the control rooms of air traffic controllers and operators in a whole physical space of interaction. The current study focused on two major communication categories, coined as silence and voice, and the knowledge they represent and provide from multiple points of view.

Silence, here, is extended as a concept to cover multiple dimensions, beyond the studies of intra-flight communication where cabin crew members fear to speak up and silence as a regulatory requirement of a sterile cockpit. This dimensionalization of silence is a clear theoretical contribution, as it specifies aspects of a conceptual and operational consideration that advance our understanding of flight realities, as social phenomena analyzed by Babbie (2009). Silence in the cockpit has the form of no-talk, not confirming/reading in checklist marking between pilot and copilot or using a silent checklist to mark. Silence includes non-replying to calls from air traffic control for extended periods, no talking between cockpit and the cabin, and standing by silent in an expected interval till the next phase of flight. At an organizational level, silence has the form of not reporting flight incidents in the flight log or in the briefing-debriefing process.

Since silence intervals are part of the interaction, vocal messages are needed to complete each interaction cycle. At this point, this book synthesizes a vocal communication channel by consolidating nonverbal communication, including paralanguage, with verbal messages (as descriptions, calls, readbacks, and hearbacks), in the category of voice.

Nonverbal is the communication without words, from vocalizations to gestures and the use of objects and artifacts (Devito 2011). Paralanguage is the vocal but nonverbal dimension of speech exchanges and has to do with how something is said and not what is said. Paralanguage is an important vocal aspect of messages describing nonverbal cues of the voice like stress, pitch, rate, volume, or mainly hesitation talk (that is rather truncated or not) and the variations of all these. Hesitation is a key aspect of paralanguage and may convey emotion or nuance meaning with forms like prosody, pitch, volume, and intonation (Vasilescou and Adda- Decker 2007). Hesitation, here, is commonly accompanied by silence as a result of uncertainty for the receiver or a reluctance to correct from the sender. Hesitation is closely related with meta-communication, as it communicates about other already transmitted messages. The complementary character (for the verbal or nonverbal message) of hesitations connects with different vocalic realizations: filling pauses, word lengthening, repetitions, and speech repairs.

Apart from that, voice (here) includes the nonverbal communication of human actors across different airplanes and cockpits, as a transition stage from silence to verbal messages. It could take the form of waiving hands across cockpits, waiving wings during flight towards another aircraft, approaching the other in proxemic distance, using territorial markers in a cockpit to signal operating status, activating silent and audible but visually mediated messages in banners and symbols from the cockpit to the passengers, or the form of radar locking in a fighter jet dogfight. All these attributes are different ways to signal a message in nonverbal channels.

Verbal messages refer to the use of words, oral and written, not to oral vocalizations. In aviation, the use of language is crucial for verbal interaction, as well as the effort to make the meaning of the interlocutors as equivalent with the meaning in words used. Along these lines, mother tongue, here, provides cultivation towards thorough understanding of the original aviation terminology in English. The use of familiar terms in a mother tongue, in specific cases, seems to facilitate the users to better understand the original aviation terms in English. It also highlights a direction for further research by investigating its role in familiarity between pilots and controllers.

Lastly, situation awareness, here, has proved to be a useful addition to communication model for proper reception and feedback, as well as meaningful feedforward for continuing interaction. SA assembles the perception of the elements in the environment, within certain time and space, with the comprehension of their meaning and the projection of their status in the near future (Endsley 1988). The discursive space of the extended cockpit, here, consists of actors that take roles and apply rules in order to complete tasks with their social actions. Conversation takes place in a time-critical situation, and all interlocutors have to keep track of responding demands. They must understand the system in its entirety, position their role in it, and remain vigilant for consequentiality. The pilot's SA means integrated understanding of factors that will contribute to the safe flying of the aircraft under normal or abnormal conditions (LA and TA, from Sect. 4.3.2). As SA increases, the pilot is increasingly able to "think ahead of the aircraft," for a wider variety of situations (a type of GA, of Sects. 2.5 and 4.3.2).

On the other hand, the air traffic controller needs to sustain GA, for the big picture ahead (in the immediate future time and for the available airspace status), while addressing/replying with LA to each different flight in the scope of approach control or air traffic control. TA is necessary for the controller to accomplish the required separation in the limited airways. SA is considered a process (to develop) and a product (of the conditions and interaction), in a certain context, at a particular point in time.

Human Factors in Aviation-Related Issues Human factors issues related to interpersonal communication have been implicated in approximately from 70% to 80% of all accidents over the past 20 years and more, especially in the issue of task management (Iani and Wickens 2007). Consequently, more than 70% of the first 28,000 reports made to NASA's ASRS (which allows pilots to confidentially report aviation incidents) were found to be related to communication problems (Connell 1995). Studies of collaboration among scientific and professional communities suggest that an initial period of physical proximity is necessary in order to build trust and to come to consensus on the focus of proposed goals and projects (Carley and Wendt 1991). The importance of understanding routine work, repeated and confirmed actions, practices, and situational requirements of the users in the design of tools and technologies that they use is recognized by many theorists (Norman and Draper 1986; Norman 1988; Winograd and Flores 1987; Moran and Anderson 1990). NASA's researchers analyzed the causes of civil aviation accidents and incidents between 1968 and 1976 (Cooper et al. 1980; Murphy 1980) and concluded that pilot error was more likely to reflect failures in team communication and coordination than deficiencies in technical proficiency. Most airline accidents are attributed to errors made by the flight crew (Loukopoulos et al. 2001, 2003, 2009).

Human factors work incorporates the holistic approach of examining the role of the human actor and human skills with technology, whereas systems analysis uses all systemic components that explain interactions, information generation/flow, and procedures. Along these lines, real-flight expert data provided in this book emphasize that the cockpit system must be tolerant to function and accomplish the mission even when diverse faults occur. Uncovering the causes of such errors is one of the greatest challenges in aviation operation and accident investigation, as many factors intermingle. Expert narratives presented in this book provide also knowledge on how to prepare for the "unexpected" with various contextual information (as explained by Kochan 2009), i.e., events not anticipated with a base probability rate or the absence of expected events. That is why a flight mission should be represented via its communication occurrences, as interactions that materialize the intangible session of a flight. This cultural and communicative point of view enriches the typical causality that searches for a *human error*, with informative interactions and oral documentation. This is the step forward to enrich the knowledge base of the field and refine the expertise level necessary for the industry.

7.2 Accident Occurrences and Airspeak Language

The modern aviation system deals with the escalation of routine flights, unusual situations, and incidents and major accidents in the taxonomy described in Sects. 1.3.2 and 3.2.

Accident Occurrences and Socioculture The current book investigates accident occurrences and draws from two landmark cases (introduced in Sects. 3.1.1 and 3.1.2) which undersign their own eponymous literature in the industry and enrich accident investigation practices.

An accident, here, as defined by the International Civil Aviation Organization (ICAO 2000, 2001), is an occurrence during a flight in which (i) a person is fatally or seriously injured, (ii) the aircraft sustains structural failure or damage, and (iii) the aircraft is completely inaccessible or missing. An *incident* is defined as an occurrence, other than an accident, associated with the operation of an aircraft which could affect the safety of operation. A *serious incident* is defined as an incident with circumstances indicating that an accident nearly occurred. Incidents are violations of instructions or legal parameters that may affect or have actually affected the safety of operations. Incidents and accidents reported in Eurocontrol repositories (Skybrary 2010) and Federal Aviation Administration (FAA) Aviation Safety Anonymous Reporting System (ASRS 2013) reveal over 60% communication problems (Tompkins 1991). Earlier studies in the 1980s indicate that those communication issues contain problematic information transfer (73% of incident data), as well as language incompatibilities rooted from socioculture and safety culture variations (Billings and Cheaney 1981). Conversation-related issues in the cockpit have been a constant knowledge concern of the international aviation community regardless of flight mission, airplane type, airplane manufacturer, operator, or geographical region (Garzone et al. 2010).

The current study touches specific issues of organizational culture (like safety and just culture), as well as topics of professional culture for pilots' habits and controller's habits that extend previous literature, like the work of Braithwaite (2001). Cockpit conversation language and socioculture affect all individual actors in overlapping layers, as shown in Fig. 7.1.

Let us go through these layers of cultures, from the inner to the outer one. Commercial, fighter pilots, controllers, and accident investigators belong to different organizations (institutions) with different fragmented cultures (the inner parallelogram). The role of hierarchy, the respective vocabulary, and the potential flexibility could be part of that organizational culture. Since they are in the aviation industry, profession, and mission, they acquire a professional culture of immediacy, living the moment and obeying rules that define the profession (both for pilots and controllers). Fighter pilots could be a smaller subset, as they are in the profession as service with duty and/or patriotic concerns and not just as occupation.

The dashed line circle overlaps with the organizational culture, as the intersection refers to aviators who exercise their own specialization in an organization.

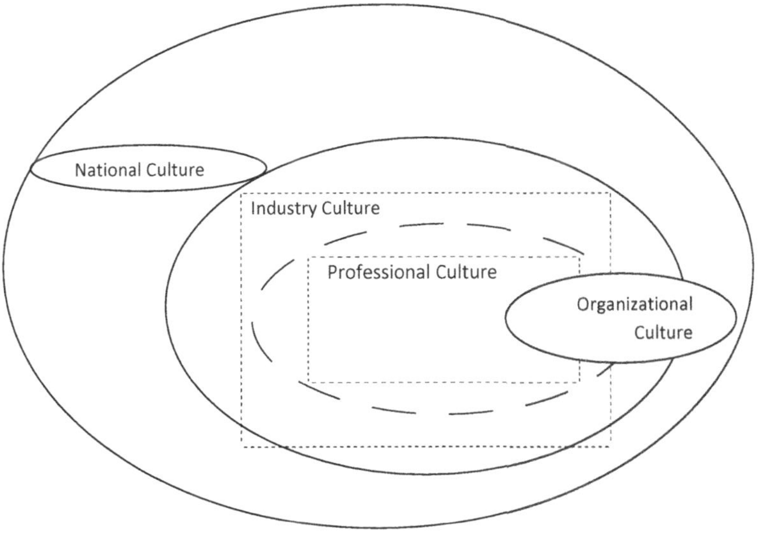

Fig. 7.1 Structures and layers of culture

Being a service industry with demanding resources, aviation has another layer in the figure, containing the organization and the profession. Promptness in time, orientation to humans as users, and cost-efficiency are attributes of this industry culture. Lastly, all participants operate in air and ground spaces of different countries or originate from other countries, where language (mother tongue) and inset may affect their meaning-making. All these layers of culture in Fig. 7.1, usually, coexist in human operators and lay the foundation for interpreting their behavior.

Airspeak English and Language Use The study for the current book highlighted the phenomenon of silence in aviation interaction, as the source of multiple semantics in meaning-making, and extends the language literature, in that sense.

Although aviation is often considered as over-proceduralized, with standard operating procedures to prevent threat and error, error is inevitable due to human limitations (Helmreich and Davies 2004). Crystal (2003) analyzed how English has become the global language over the past 50 years, mainly due to the power status of its native speakers and not to their number. Global English became a communication platform for business, science, and technology. It is a platform beyond a language of special purposes or globish English (Nerrière 2004). Furthermore, Crystal recognized conversation registers such as "seaspeak" for maritime use and "airspeak" for air traffic control. On the other hand, the need for bilingual ATC communications was underlined in empirical cases with aviation professionals. Ragan (1994) suggests that "airspeak" is a controlled language in aviation that is broadly regarded as acultural; in contrast, cultural factors and conversation interaction do have an effect on specialized communication. Conversation language has organization, and meaning in everyday-life activities is shaped by context and renews the context, as well. Order is maintained during interaction, and routine conversation lays the ground for the structure and organization of institutional talk (Person 1999). Airspeak conversation is a controlled language with a singular context focusing on human-machine interaction and air transport safety conduct.

In this sense, cockpit acts of conversation and airspeak describe the special character of cockpit conversations with internal (on board) participants and the external ones, most notably the air traffic controllers; all participants are considered actors who follow the basic model of communication. Additional supporting literature on speech acts, actor-network theory, trust and trust in automation, as well as rules and roles and identity in the workplace informs the discussion of the current book.

The way language is used is a known source of misunderstandings in stressful situations. Instances like false interpretations, filling out incomplete statements in ways not intended by the speaker, and holding back important information for reasons of politeness and face-saving have been documented repeatedly in actions like fire fighter operations and disaster relief missions (Ungerer 2004). Ambiguity was involved in the most severe accident in commercial aviation history, the (ground) collision of two airplanes at the Tenerife (Norte Los Rodeos) Airport in March 27, 1977. This incident centers around the question of whether the phrase *we are now at takeoff* is to be interpreted as *we are now at the takeoff point* or as a kind of continuous tense, *we are [being] now in the process of taking off* (Roitsch et al. 1978).

There are other such word ambiguities or structural ambiguities in aviation language. Cushing (1994) mentions the verb *hold*, which in aviation terminology means "stop what you are doing right now," but in ordinary English can also mean "continue what you are doing right now." Linguistic investigations of cockpit voice recorder have documented the effect of linguistic misunderstandings on plane crashes (MacPherson 1998; Cushing 1988, 1994, 1995, 1997). From the point of view of aviation professionals, a flight is accomplished through group work, and small groups provide an ideal environment for developing linguistic security, a notion important for aviation safety (DeMatteis 2011). Linguistic security is realized among cockpit interlocutors when they see their language as acceptable (*from those with whom one is most concerned about communicating*; Koch 1975). This sense of security is a greater challenge when human actors in the cockpit have mother tongues other than English, as language is a fundamental marker of their identity (King 1997).

That is why the current book proposes the phrase *cockpit knowledge acts of conversation and trust* to incorporate all relevant conversation aspects in that conversation-scape.

7.3 Silence-Related Theoretical Contributions

The concept of silence, in this book, is explored as used and understood from several aviation professionals on the base of real scenarios or in the context of their work. When pilots talk to each other or to air traffic controllers, there are periods of silence or no-talk, sometimes substantially longer than in ordinary everyday-life conversations. There are particular stages, like the cruising phase with the autopilot, when silence is measured for many seconds or many minutes. Non-talk activities from pilots are the use of cockpit controls, displays, and written materials in charts and checklists, as well as diagnostic activities to identify unusual situations. Talk and non-talk activities begin and end, and they are relative to repeated non-talk periods which are coordinated during the flight. That means that silence flows in the conversation patterns used during flight communication and has an interactional effect for the work of both pilots and air traffic controllers (who have to follow these periods for multiple flights in each of their shifts).

The responders laid their framework on silence using their own recollections of typical examples or examples marked in their mind and are determinants of their decision-making. Even in case of accidents, the crews and the controllers conducted their flights as routine, until the moment when things start to go wrong. Let us go over the results from analyzing the responders' insights.

Organizational reporting and scapegoating process were pointed out in the Falcon VIP flight, as a view of human error. Silence was a latent phenomenon before takeoff, during the critical 24 s of descent, and a crucial factor in the accident analysis of cockpit recordings in the Falcon VIP flight, as described in Chap. 3 (and further). *Ifaistos Limnios* introduced a landmark case of the "shaking VIP flight," in which silence had a deeper content in organizational reporting: to speak up regardless how many times you are not heard and address the appropriate levels responsible for accountability. Also, in this case, the concept of scapegoating for human error emerged, with its consequences in aviation safety, as well as in the complexity of civil aviation with economic, market, and political concerns.

Silence was a dominant phenomenon in the Helios 522 ghost plane, which *Megas Alexandros* described as an outsider, from the accident investigation side, in Chap. 3. It may be one of the handpicked examples in which the hindsight mechanism (there are no survivors and very limited data from the recordings exist) is necessary to communicate the silence of a haunted story. Helios 522 flight redefines what it means to communicate in a timely manner, without waiting silent for feedback. Besides that, it started the discussion for the open channels of communication between the cockpit and the cabin crew members. The Falcon VIP flight and the Helios 522 flight constructed a system of reference and knowledge creation with consequences, regarding silence for all professionals who were on duty when the accidents occurred.

Air traffic controller *Ikaria Ikaros* brought up a categorization which prioritizes in-flight activities and no-talk periods with distress or emergency situations, while the controller is waiting for feedback from the pilot who may need assistance. There are instances when silence is the magic word to keep, she emphasized. Also, she pointed out the importance of corporate brand, the position of organizational culture in the brand, and the lack of motivation for crew members to frankly speak up.

Oia Santorinis, another air traffic controller, emphasized that miscommunication with a pilot may result in loss of SA, if not repaired by the copilot who had situation awareness (LA and probably TA type) and copied the received clearance but remained with silence. She pointed the informal grapevine microculture of humiliation which interferes with safety culture of the airliner, as well as with not well-developed just culture.

Mutual silence with the implicit personality emerges when moving on to fighter pilots' culture and expertise. *Eleftherios Venizelos* discussed the relation between trainee (fighter) pilot and instructor officer. It is a relation of closeness and mentoring which may have a downside. Trust is reset (and reinforced) when the instructor admits his mistake to the trainee. *Ioannis Daskalogiannis*, another fighter pilot, emphasized the institutional response. He connected silence with SA and the leadership style necessary to lead by example. The person, who speaks in a flight, is the one who is aware of the situation. Rank does not work in the long run, and operational respect is gained when you fly with the others, he emphasized.

Lastly, *Ippokratis Koos*, a helicopter pilot connects with operating barriers. Marking a checklist or observing the checklist automation is decisive for flight cooperation between pilot-in-command and copilot. Also, silence could be a result of perceived conflict between efficiency/passenger comfort and safety. Silence could block reporting of all flight parameters. He also felt the impact of silence from his flight engineer to reinforce an open communication policy for operational issues in his briefings, before the final word of his mission command. So, silence is strongly related to the culture of authority or collaboration and affects flight and mission operating parameters.

Several cases of accident flights showed that failure to speak up or report appropriately can have devastating consequences. Silence is taking turns with interaction in emergency situations to protect the peace of mind necessary to deal with the mission critical environment, human lives, and integrity of equipment. Interaction is effective when information elaboration occurs. The compulsion to communicate is an urge which harmonizes the actions of individual actors in the interests of the other members of the aviation community, at large. Thus, aviators and controllers agree to proceed or take the next speaking turn from a period of silence: to receive something that pilots say, something that pilots say or do, or something about how or why they say or do it. Long silences could indicate an escalated problematic situation, whereas no silence could flood communication channels. Voice is a complementary channel that succeeds silence, and one takes turns from the other, in normal flight conditions.

7.4 Voice- and Language-Related Theoretical Contributions

Chapter 6 was the transition from the phenomenon of silence in Chap. 5 to the use of nonverbal communication, hesitation, and verbal in aviation communication. Voice-categorical label connects silence with paralanguage and mediates the transition from nonverbal to verbal communication and the respective knowledge produced.

Reconnaissance, interception, nonverbal gestures, and verbal messages communicated the Renegade process of the Helios 522 flight, with *Neos Aghialos* as the leading pilot. The two scrambled fighter planes took off in a mission that started from reconnaissance and continued to interception, without the final phase (of shooting down), as Helios plane crashed by itself. The Renegade response protocol is described in terms of its communication significance. Furthermore, the agonizing effort of using the fighter plane as a communication tool (by waving the wings), the cockpit-to-cockpit face-to-face communication, the hand gestures, the affordances of time, the verbal communication of the emergency in two levels (inside Helios flight to the cockpit voice recorder and the flight data recorder and outside from the F-16 to the receiving control centers), as well as the moment of breath and the moral inquiries of the pilot establish a unique case in aviation communication.

This flight description is communicated by out-flight real-time observation conducted by the interception fighter planes' pair, since no direct communication was received (by out-of-airplane receivers, except the flight recorders located inside the plane) from the silent aircraft. Verbal messages from the scrambled fighter pair meta-communicated the situation when original communication was continuous silence. The commentary of Megas Alexandros, the accident investigator, provided unknown insights regarding the Renegade protocol communication and the moral basis of the shooting-down dilemmas from the authorized actors.

In the scenario of hesitation, an aircraft with a takeoff clearance received a truncated tower call in a hesitant voice. Prudence was the strategy to cope with. The pilot decided to abort the takeoff, as they were in the rolling phase and uncertainty captured their mind of what the controller's concern was. The pilot and copilot confirmed the prudential action to bypass the clearance and cancel the takeoff at the same moment of which the controller prompted them to disregard his truncated message with no other explanation. This scenario is described as if they were doing debriefing, by evaluating the action of a pilot after it happened.

Eleftherios Venizelos, the fighter pilot, argued that hesitation leads to uncertainty and misinterpretations. Safety is the primary concern when violating a clearance, in comparison with any other criteria that may apply (corporate, organizational, scheduling, etc.). Controllers' messages should be immediate and clear to avoid uncertainty and misinterpretation risk.

Then, Diagoras Rodios, a civil aviation pilot, argued on the pragmatic basis of a bad moment (a random factor) and the captain's profile. Pilot's interpretation depends on a multitude of factors which assemble to "triage," the skill of prioritizing by abstracting information which originates from the medical field. Along these lines, the flying instinct (the Langewiesche's "buoyancy" of Sect. 6.4.2) could be a differentiating factor to cope with hesitation. Then, Ikaria Ikaros, the most experienced controller, argued on the connection of hesitation with fear and situation awareness. The pilot has taken a prudential decision, given the circumstances, and Ikaria Ikaros added the fear of the air traffic controller, with the recording device in the background, and the recent trend of litigations of aviation professional in incident or accident cases. The content and style of verbal communication may depend on the availability of a recording device (the recording transcript might be considered a factual knowledge base *of what we know and not always of what happened*). Another explanation, for hesitation, is the loss of situation awareness from the controller at the time.

The next discussant, Oia Santorinis another air traffic controller, argued for the collaborative environment that is necessary for pilots and controllers to feel (together) in the same group. The controller should work as the alter ego of the pilot, providing the "eyes and the ears" for the ground when the pilot is in the air (or for the surroundings, when the airplane is on the ground, closer to GA type). Moreover, she offered a well-versed suggestion on the human error. Human error should be accepted as a reality. An erroneous instruction with a stable voice is preferred from hesitation, and uncertainty should be explicit. In the next turn of the communicative transaction, a clear correction could clarify the situation and restore the preestablished trust in a fiduciary relationship. It is a constructive view of human error with the required transparent audit trail; a type of knowledge that starts from understanding error, from a realistic point of you, and building on trial-and-error training scenarios in an evolving process.

Then, Chap. 6 introduces the scenario of language directness, informality, and familiarity by Stan Dardman, a civil aviation pilot, who speaks about the ways of approaching the relationship with air traffic controllers. This scenario continues with the addition of the importance of understanding the "other part" (pilots the controllers and controllers the pilots) and their personality traits. Stan Dardman is adding that it took him more than 1000 h of flights in order to start feeling comfortable with a level of knowledge that would allow him to interpret rules and situations with his tacit knowledge, after applying the explicit one from his training. The question that remains open for research is the following: *Is there a volume of knowledge that allows rules' interpretation and what is the criterion to reach that knowledge level*?

Oia Santorinis is the controller to reply in this scenario, citing the importance of professional ethics as a controller and the effort she makes to understand the profile of pilots better, when they do their work (even observe them in the cockpit when they aviate). Eleftherios Venizelos is the fighter pilot who replies arguing on the importance of standardized communication but also the need of constructing a social and collaborative operating environment with the air traffic controllers. And he has been experiencing that for 30 years in his career; so, it's not a theoretical ideal, it is real. Ippokratis Koos, the helicopter pilot from the Army, is the last responder. He identifies the category of interaction in the home base familiar environment with an accepted deviation from standardization, which is sometimes instrumental for routine functions. He also, acknowledges the generic category of keeping the standardized phraseology, in other missions and non-familiar environments to avoid miscommunication that may be fatal. Ippokratis Koos underlines a case of mindfulness during flights, when an expert performer pilot develops the necessary self-monitoring.

In the last part of Chap.6, after an overview of language and English use in aviation communication, an additional scenario is introduced to lay the foundation for future research in pilot-controller communication; the mother tongue as a primary code-switching element. The modern aviation system requires the maximum amount of information expressed with the minimum of effort and consuming the minimum of time. This is often addressed as Macdonaldization of language.

Palios Ellinikos is a pilot in HCAA Flight School, who shares his problem on the lack of standardization of the spoken English language and the airspeak English and the lack of conformity between FAA English expressions and ICAO English expressions. These discrepancies make it difficult to localize aviation English to a standard terminology of Greek (as mother tongue for domestic flights). Palios Ellinikos shares his practice of going to mother tongue out of routine flights and to mix Greek and English when going over checklists to avoid automated marking. The responses in this scenario show one direction of further research initiated with the Special Committee of Air Traffic Management Terminology (OMEODEK, in Greek language acronym). OMEODEK was established after I have submitted an ad hoc institutional proposal to HCAA and the Hellenic Society for Terminology (ELETO). ELETO is the scientific society for investigating and promoting neologisms standardization and the use of language in a national context.

Controllers Ikaria Ikaros and Oia Santorinis and fighter/helicopter pilots Eleftherios Venizelos and Ippokratis Koos emphasize the distinction between routine and emergency flights/situations. In the latter case, the mother tongue is the selection in all cases when an emergency occurs or an extraordinary situation is in progress (e.g., a dogfight is developing, a mechanical damage occurs, or severe weather presents a threat). The persistence in the use of mother tongue, even blended with a few English aviation jargons, has traces of the linguistic relativism notion. It seems that mother tongue, and Greek language in particular which exists in a continuum since antiquity (offering strong roots in terminology and epistemology of global sciences and humanities), is a language shaping several aspects of the professional and institutional subcultures in aviation (and for the most part in the Air Force). The connection of emotional reactions under pressure and language speaking falls in the same dimension, as well as an introduction of a concept of affordances for the mother tongue (i.e., *Does the mother tongue allow more or better communication with its tools than standardized English language*?).

Lastly, the importance of language localization for ICAO standards as a major training tool and a mental exercise of translatology is already another inquiry during the earlier research stages for the current book (Katerinakis et al. 2013). An airplane flight is a communication session in the high-complexity environment of the flight deck during the interaction of air traffic controllers and pilots following the model: *spoken-heard-understood-applied*. The criticality of language use in aviation safety is based on linguistic phenomena like vocalization, homophony, punctuation, ambiguity, inference, and repetition. Airspeak Greek, analogically to airspeak English in the mind of the responders, could construct a concept of language security and trustworthiness for national public policy in the Greek (national) airspace.

7.5 Situation Awareness Communication-Related Contributions

Both conceptual categories of silence and voice have a significant impact on situation awareness (SA) of all the actors in the GT approach used to represent knowledge (as analyzed in Chap. 4, mainly Sect. 4.3). Situation awareness is a causal category with a crucial effect on aviation safety. In review of Sect. 4.3.2 and several empirical data from the responders, SA emerges as a useful addition to the communication model of interaction with perception, comprehension, and projection, as its primary functions.

SA was formally introduced in human factors, critical systems design, and cognitive engineering literature by the empirical domain of war tactics (with monitoring the opponent with slow processes) and sports management (with an implicit immediacy, as situational awareness). During World War I, the identification of SA came with the realization of the importance of being aware of the enemy before the enemy does so, in order to communicate the appropriate commands and act accordingly (after the commander "knows" and "order"). SA distinguishes the understanding of system status by human operators and the actual system status itself. Thus, it differentiates the users who interact from the channel and the context they use. Knowledge negotiated and actions prioritized in the process follow consciousness of the actors and common knowledge principles, in the knowledge framework explained in Kvanvig (2006).

SA covers the individual's perception of the "here-and-now" while this state adjusts in time. SA includes perceiving the elements of the environment (who is where, in the interaction), comprehending their meaning (what is the exchange content and what is the initiation), and projecting their status in the future (what they will do for feedback or completion of the interaction). The three types of SA – the local awareness (LA), the transitory awareness (TA), and the global awareness (GA) – reflect communicative interactions better, as they include the state of the individual actor, the cooperating actors, and the time dimension (as showed in Fig. 4.1 of Sect. 4.3.2). By dividing SA in parts, information processing tasks are added prior to the performed tasks. The result would be a better prepared message to continue interaction and, consequently, task execution.

The complete picture of SA is synthetic combination of different viewpoints. The functions of perception, comprehension, and projection are sequential and parallel, and "awareness" starts at a local level (as current knowledge of task-related events), continues at a transitory level (as mere awareness moving from one situation to the other with turn-taking in communication), and becomes global (as assessment of the full scope picture of flying, controlling, or maintaining an aircraft). The reply to "who is where," for example, includes actors, objects, and features of the environment of the situation. Awareness is a variable state of knowing adjustment and an attainment from the answers in the three questions: who is where, what is said, and how the state is projected in the near future. In this model, the three questions are considered to be components in the intersection of "what we know." Each component is a stage that requires some data from different forms of information processing. Data input, processing, and reasoning all overlap (as discrete items and as a continuous flow) to lead to SA.

In dynamic interactions the situation is in fluctuation, and all data, evaluations, and assessments are of a transitory nature. Air traffic control and airfield ground movement of aircrafts, as well as most forms of football, basketball, or disaster management deployment are activities to apply this model. The combined answers to the three questions, above, formulate the transitory awareness (TA).

Two other types of awareness complement the viewpoints of the concept; the local awareness (LA) and the global awareness (GA). The decisive factors for the separation for these viewpoints are the constrained time of limited availability, the span of command (the number of people who report/communicate to a commander rather than the commander's rank), and the geographical area of interest (or the medium, like air navigation or sea cruising), with changing conditions. The message follows a path of a slower process for storage (as memory) and information processing, so that "awareness" would embrace locally situated and then global events (to evolve to "knowing").

A typical example for that communication flow with SA would be a flight with commercial pilots in a two-person crew. Each pilot has a different LA. The (flying) pilot-in-command who is the pilot will have inputs from the controls (and the actions performed) which may not be available to the copilot. The rate of change for cues in the environment and the key items of information determine measures of awareness. A military pilot in a ground attack mission experiences a far greater change to the TA than a commercial pilot in the final approach to land, although the LA situation may be quite similar. Thus, indicators that are of immediate importance in TA may decline in importance to contribute to LA. Variations in data importance are associated with the changes in operating and flying environment, as an ecological perspective.

When SA is conclusive for more actors involved, then the typology of TA, LA, and GA is closer to reality, from multilateral points of view. In this sense, the air traffic approach controller and the pilot of an aircraft on final approach have a common goal and they are both concerned with the same high-level task, but they attend different data inputs and acquire different degrees of TA, LA, and GA. Furthermore, when the aircraft is moving on the ground (like in Sects. 6.4.3 or 6.4.4) TA, LA, and GA measures change again.

In the example of Helios 522 flight, the two fighter pilots intercepting the silent passenger aircraft possess TA (after their briefing) and LA but low GA, where as the Chief of HAF (in command post) had a high TA. The Defense Minister and the Prime Minister had high GA but limited TA and lack of LA. The pilot flying in the interception communicates his LA to increase the TA of the Chief of HAF, who in turn conveys the messages to the Prime Minister who has to rely on the communication flow of lower-ranked actors and defend the necessity with his decision.

In the typology of awareness, the transitory type is closer to the group communication and teamwork. The local type fits typical perception in communication models. Also, the local and global types are linked with situations of multiple levels of hierarchy and intraorganizational communication. The separation of SA in types denotes that an individual could commit in awareness of any type (or of all types at once). Real-life topical factors and message flow (in flight, combat, or other activity) indicate which type is relevant. As it happens in communication patterns, the *participants' confidence* in their situation awareness might be different than their *actual situation awareness*. Receivers' perception of a message may be different than the intended content of the message.

SA, especially in the expansion of the three types, implies clear communicative dimensions as human actors should act upon, respond to, and confirm/correct important informational cues. Apart from that, human communication forwards the information flow functioning in LA type and facilitates the move to TA type. The GA type needs also reporting and critical evaluation and access to privileged operational and mission information. Awareness (of all types) is internal to the individual and the meaning-making process and does not reside on any (technological) information display. The individual actor develops awareness when considering the interpersonal interactions and relationships with relevant communication. Interpersonal communication, here, includes the form of the "individual-group-team", or the form of "own vs opposing force/group" in command and control contexts.

SA Realities in Flight Communication The high level of standardization in aviation is expected to streamline SA, but the operating environment still has humans with multiple languages and mindsets working together and interacting in international flights (in diverse conditions).

The deadly Tenerife disaster at Los Rodeos Airport on 1977 (described in Sect. 4.3.2) is a clear example of the effect of language, meaning, and its context to LA and TA types of awareness. The Dutch copilot being in the takeoff process was not understood by the Spanish controller who replied with a vague "OK." The Dutch captain mistook the meaning of "clearance" at that specific point in time and powered up, replying with a hanging "Let's go." Then the controller called for "standby" and increased complexity in the interaction. A dynamic flux in TA and LA types from one utterance to the other occurred in the 8 min before the collision, as the controller had a change to the GA type due to the ground movement of the US flight. The controller spoke simultaneously with the US flight crew and the crew members in the Dutch flight find it hard to hear. Neither the copilot nor the flight engineer, in the Dutch flight, questioned their pilot-in-command who had a senior position (as it happens in several responses in Chaps. 5 and 6), and the impact occurred about 13 s later.

In the Tenerife disaster, derangement of all types of awareness for the different actors was combined with additional factors (fog, small and unfamiliar aerodrome, news distractions, and the expedition and rush to cover the schedule delays from previous days). Information storage and processing for all these factors were very important for appropriate message transmission in the communication among the control tower and the two aircrafts. The information parameters of SA are valuable inputs for communication, and consequently, knowledgeable processes.

In rescue, disaster relief, and military coalition operations, the cooperation of multinational and multi-force teams increased the potential for misunderstanding due to incomplete SA and the relative use of different languages. The relay of commands and the confirmation of information flow are a demanding communication accomplishment which has an effect to awareness of LA, TA, and GA type. A similar situation of interpretive messages, fragmented information flow, and gradual development of awareness happens in NATO reconnaissance flights with multination crews of the NATO AWACS (like the pilot who responds in Sect. 5.5.1). These are instances of major communication challenges, in terms of language and culture, which clearly affect SA (and consequently effectiveness and safety) especially in the immediacy of a flight situation.

In the emerging causal category of SA in this study, four key elements cooperate: humans (as users and interactants), informational cues (as stimuli processed), behavioral cues (paralanguage and role expectation attributes), and appropriateness (as regularity) of the responses. This regularity implies the comparison of the actual response with an anticipated response (a choice from a number of possible expected responses). These expected responses formulate an instrument for a performance measurement of SA, from both pilots and controller's sides, as it is apparent in the responses analyzed in Chaps. 5 and 6. That measurement constructs the knowledge level we acquire, as well as the operational knowledge applied in action.

For example, for a HAF pilot, SA (TA type) means to understand the threats and intentions of the "non-friendly" aircraft or enemy forces, as well as the status of his own aircraft (LA type) and communicate that way. In order to project the situation on his near future, it is necessary to communicate from the start (communication builds a foundation for a GA type also). For an air traffic controller, SA requires estimation about current aircraft positions and use of flight plans to arrange the necessary separation between aircraft positions and entities in the airspace (GA type), in order to predict future states of potential conflict (LA type). Of course, the controller is not an observer and needs to communicate in all stages and has a considerable effect to the SA of the cockpit crew (TA type for the controller).

Diagoras Rodios's suggestion who thinks that triage and allocation of resources are priority, we found an occupational skill that should be part of situation awareness response for the pilot. The context determines, to a large extent, the meaning (as negotiated and constructed) of verbal and nonverbal behaviors. The way that pilots perceive incoming signals demands the abstract thinking, immediacy, and priority handling of triage to select and prioritize the next actions, in Diagoras Rodios's view.

Thus, as an emerging communicative parameter, situation awareness includes acquisition of information from the immediate (and the co-present controller) surroundings, integration with relevant knowledge of the situation, exploration of a mental picture for the situation, and anticipation of reactions for the immediate future. Hence, situation awareness has a clear effect on aviation safety and should become an added component in communication models. It is a direct contribution in knowledge creation as mutually constructed between actors and interlocutors; they were able to act in real time, and their knowledge depository combines tacit, skill, and procedural elements.

7.6 Discussion and Partnerships with Practitioners

I have conceived the current book as a research project in the basis of human communication in highly mediated environments, safety, language security (standardization and propagation of mother tongue terminology with aviation English), pilot knowledge and expert performance, and a constructive approach to human error in organizational behavior. Analysis for data collected during the course of this ongoing project resulted to a doctoral dissertation which focused on the concept of "silence and voice" (as labels of categories) and situation awareness, as a contributor to aviation communication for safety. But several other topics emerge en route and could develop further beyond with research orientation.

Collaboration with Aviation Communities In the early stages of this research, I became engaged to the work of HAAISB, Hellenic Aviation Incident Reporting Committee (EAPA), the HCAA Flight School (SPOA), the daily routine of Hellenic Air Force Squadrons, and the challenges of work in control towers. During these interactions, I have proposed to upgrade language and terminology training, on the basis of mother tongue, in order to position Greek aviation English in the appropriate terminological level, as a field of study in the Hellenic Society for Terminology (ELETO). In a cooperative effort with the SPOA School for air traffic control, a new permanent committee on Air Traffic Management Terminology (OMEODEK) was institutionalized inside the HCAA (Papadimitropoulou 2012). The scientific foundation and support for this committee derive from ELETO,[1] a terminological society with strong tradition in working on neologism and translatology, with my participation as well (this is an institutional and scientific partnership). The initial contacts for this study, as well as the discussion with several of the responders started in contacts of 2011. The first outcome of this collaboration was the paper on aviation terminology, which presented the topic for the first time in the map of scientific inquiry beyond what practitioners do, in the Greek scientific community, and its specialized society on Greek terminology (Katerinakis et al. 2013). These activities generated a substantial motivation for the community of air traffic controllers and aviators in Greece, who have now created an informal discussion network to follow through the linguistic and standardization evolutions and relevant consultations.[2] Similar research and practitioners' partnerships could develop in diverse national cultures.

---

[1]ELETO presented this new collaboration in the Proceedings of the Athens Delegation of the European Commission on "A Terminological Network and Translation" (April 11, 2013) at http://ec.europa.eu/greece/pdf/2014-06-03-praktika-imeridas+parartimata.pdf. Also in each ELETO's Annual Report of 2013 at http://www.eleto.gr/download/OtherEvents/2014-06-11_GA2014_GESY-Report_EN.pdf.

[2]A Facebook group (at https://www.facebook.com/groups/OMEODEK/) administered by ATCs Kostas Patouras and Nikos Papadopoulos is running to accommodate the discussion about aviation terminology. This platform integrates the contributions of members of the aviation community, translators, practitioners, scientists, and regulators.

The work with OMEODEK and the resultant inquiry had already an impact to curriculum adjustments in SPOA and in language proficiency evaluations. The mother tongue (widely used in domestic flights, in the dense network of more than 40 airports in Greece) could be a useful tool to better understand the standardized ICAO phraseology.

EAPA committee of anonymous reporting was rather introvert in operation. My application was the first one, in their records, with an inquiry for data (2012). Now, it works regularly in using standardized incident monitoring at a national level and, for EASA, at a European level. Similarly, the HAAISB received the first two inquiries for research in communication issues with my application (2012–2013). Those inquiries contributed in generating an atmosphere of accountability to the public, as both committees have a significant role in the development of safety culture in the sector.

Further Research I am not approaching aviation realities as a tabula rasa, as there are several preconceptions about the field of aviation, its position in social life of individual actors, and its role in collective society (at least in the major case of Greece). My personal background in the Air Force, the bonding cultivated in the time of service in real crises, and the affiliation with several businesses active in military procurement due to my business involvement provide familiarity, access to people and sites, entrusted discussions, and valuation criteria. The work on aviation terminology and safety, as well as the study of the military defensive culture and rule-based knowledge and expertise are clear research paths for the future. An additional challenge, due to the role of aviation in Greek society, is to investigate the roots of effectiveness and organizational performance that we meet in the Air Force in order to apply them in other managerial and institutional structures.

The ground for further research is laid with the combination of further inquiry in scenarios and correlations that seem plausible in the operating culture of Air Force and the service culture of aviation. Also, the deeper investigation of the decision-making conditions from pilots in crises is expected to be a fruitful path. In the case of air traffic controllers, their co-presence should be employed in any cockpit-related interaction to broaden the scope of analysis.

Further investigation of the issue of mother tongue in aviation communication could cover in two dimensions: as a channel of cultivation to better understand the instances of the respective aviation English terminology and as the preferred tool in emergency situations. The desired clarity of *one-word one-meaning* is a certain direction in the latter dimension.

At an institutional level, the engagement in standardization localization processes at EU or broader geographical level should include communication. One of the keystones of European integration is the localization of institutional and regulatory content in the language of member states. What is the trade-off between a lingua franca and a mother tongue, along these lines? The same question is applicable in various international contexts.

“Triage” is a concise notion, a common practice in the medical field that seems to have analogies in the field of aviation. The abstract idea of mindfulness and the skill to prioritize activities and critically extract information from a pressing situation could find an inclusive combination using triage.

Fighter pilots seem to advance in expert performance and self-monitoring skills. When they operate in a defensive military doctrine, the reactions and communication patterns seem to be different in comparison to those of offensive missions. Also, their interpretation of tacit knowledge needs further inquiry regarding the balance between mission commands and their flying efficiency.

The briefing and debriefing process is in the heart of managerial efficiency of the sector. That process of communication and meta-communication could provide valuable input in other organizational settings.

The use of checklists as communication tools is an under-investigated topic that could provide useful insights for quality assurance purposes. Also, the checklist format, as message design configuration, is a clear communicative dimension that should be combined with human communication capacity. The tendency of checklist automation and over-reliance in automation sometimes equals to deskilling. It is a phenomenon that needs further investigation.

The current study highlighted the importance of human communication, with the categories of silence (personal, operational, institutional, and regulatory) and voice (hesitation, nonverbal and verbal communication) and the concept of situation awareness in order to accomplish a safe flight mission. Little was known, before, about the human communication use of silence and voice in flights and their impact on effective flight communication, especially in unusual or emergency situations. Also, the flight pattern of “aviate, navigate, communicate” underlines the category of situation awareness. The inquiry on the real-life scenarios used for this study will continue in search for the aviation instinct.

7.7 Overarching Questions Still Fly

The study for this book brings to light the importance of communication, via active conversation, in order to accomplish a flight mission and the necessary knowledge applied, according to those who execute the flight (air and ground partners). Conversation exchanges in flights are real-time speech acts; participants must do what they communicate, but, even when they just report facts, they meta-communicate confirmations of completed or attempted actions, as they do then. Standard operating procedures and rules of engagement restrain the language used, while spontaneous interaction may deviate from standards in style or content (English or other, with code switching).

An aircraft flight represented in the basic model of communication is an ideal flight “session acts” situation. Interlocutors work together to ensure that their utterances are mutually understood, as pilots are “many” to address “one controller” with accuracy and in a rapid pace. Controller-pilot communication follows a collaborative scheme to conclude with the desired result. The discursive space of a cockpit includes participants with tasks, procedural and organizational rules, mediating devices and affordances for social interaction as abilities to manage social interactions, emotional context, or social presence. The cockpit is a machine-supported collaborative environment in which attention management (cross-device interaction, cross-mode interaction), awareness-showing, context management, relationship building, and maintenance are instrumental. Human actors must understand the system, position their role in it, interpret and apply SOPs and RoE, and remain conscious of "what makes the difference" when they deviate from "what is expected."

In this context, there are several communication-oriented problems that arise. One set of problems involves issues of *who is addressing whom and how feedback loop is maintained.* A second set of problems concerns the actual practice of SOPs/ RoE and what counts as following them in a crisis. The third overarching factor is *how those processes need to be negotiated in time-critical situations.* In this study we examine conversation acts in the cockpit with ATC, application of rules, registers and protocols, and managing the affordances of time in flight instances. Also, in the process, we investigate how interlocutors sustain trusting identities and culture in crisis warnings and conversation.

Several key findings broaden the strategic inquires rooted in the current study to new knowledgeable dimensions. Understanding becomes knowing via situation awareness (LA, TA, GA) that seems to be an enriching addition to information models. Hesitation is an early symptom and action when situation awareness lacks, and redundancy can be a message amplifier in several communication contexts. Nonverbal communication, in institutional settings and mediated environments, is instrumental (produces operational speech acts). Silence is an interactional element with substantial contribution both to completed flights (when needed) and aviation accidents (when it distorts the message). Human errors should be acknowledged, talked, and repaired with transparent communication. Human error repair (not excuse) evolves as a systemic feedback in communication and relationship building for mission critical environments. Simultaneously, systemic feedback cultivates institutional knowledge and operational memory in organizational settings. Safety culture and just culture demand continuous intra-institutional communication (not only outreach). Deontological inquiries and moral judgments should always be part of message evaluation and knowledgeable action in communication models. Local rationality, founded in the "here and now" of the flight, is a condition to evaluate communicative behaviors and knowledgeability. Aviation subcultures contain several distinct microcultures which affect professional responsibility, uncertainty avoidance response, and decision-making in microenvironments. Tacit knowledge of pilots, controllers, and accident investigators is a necessary addition to SOPs and RoE and should be exercised and monitored.

Lastly, at a policy level, innovative investigation challenges are expected to augment knowledge accumulation regarding flights and mission critical environments:

(i) Mother tongue preference or last resort medium in critical situations.
(ii) Standardization of English lingua franca with localization criteria and the culture of the discipline.
(iii) Triage, as abstract mindfulness and as a performance skill.
(iv) The expertise and expert performance in the Air Force towards the manuals.
(v) Managerial efficiency through briefing-debriefing processes.
(vi) Profile differences of former military pilots in civil aviation vs the originally employed commercial aviators.
(vii) Communicative utility and power of checklists, as applied knowledge and knowledge to be designed.
(viii) Human consequences for pilots saved from a crash and the loss of the desire to fly.
(ix) The culture of defensive military doctrines when armed forces do not attack vs the war-machine attack culture.

Flights go on and accidents pass through. Inquiry is still there, and the demand of in-depth knowledge is broadening, as critical environments increase in number and relevant situations are multiplied. Paraphrasing Amelia Earhart's saying,[3] in the end "we are all passengers." This book aims to put the reader in the cruising position, as the pilot-in-command with the empirical data available. Knowledge comes with awareness of the criticality at stake, starting from the flying instinct.

References

ASRS (Aviation Safety Reporting System). (2013). What would you have done? *Callback* 396 (situation #4). NASA Aviation Safety Reporting System, January 2013. Available at http://asrs.arc.nasa.gov/publications/callback/cb_396.html

Babbie, E. (2009). *The practice of social research* (13th ed.). Wadsworth: Cengage Learning.

Barshi, I. (1998). *The effects of mental representation on performance in a navigation task.* Ph.D. dissertation in Cognitive Psychology, University of Colorado, Boulder.

Billings, C. E., & Cheaney, E. S. (1981). *Information transfer problems in the aviation system* (NASA TP 1875). Moffet Field: NASA- Ames Research Center.

Braithwaite, G. R. (2001). *Attitude or latitude? Australian aviation safety*. Farnham: Ashgate Publishing.

Carley, K., & Wendt, K. (1991). Electronic mail and scientific communication. *Knowledge, 12*(4), 406–440.

Checkland, P. (2006). *Systems thinking, systems practice* (9th ed.). London: Wiley.

Connell, L. (1995). Pilot and controller communications issues. In B. G. Kanki & O. V. Prinzo (Eds.), *Proceedings of the methods & metrics of voice communication workshop*. Washington, DC: FAA.

[3]Amelia Earhart was an American aviator (1897–1937), who has set many flying records, as the first woman to fly solo across the Atlantic Ocean, across the USA, and in a roundtrip to/from Hawaii. She has received the US Congress Distinguished Flying Cross for "extraordinary achievement in an aerial flight." She disappeared in flight (in an attempt for an eastbound flight around the world); no wreckage was found (mysteriously) and was declared "lost at sea."

Cooper, G. E., White, M. D., & Lauber, J. K. (Eds.). (1980). *Resource management on the flightdeck: Proceedings of a NASA/industry workshop (NASA CP-2120)*. Moffett Field: NASA-Ames Research Center.

Crystal, D. (2003). *English as a global language* (2nd ed. 1st ed., 1997). Cambridge: Cambridge University Press.

Cushing, S. (1988). *Language and communication-related problems of aviation safety*. Washington, DC: U.S. Department of Education (ERIC document reproduction service no. ED 296595, FL 017 504.).

Cushing, S. (1994). *Fatal words: Communication clashes and aircraft crashes*. Chicago: The University of Chicago Press.

Cushing, S. (1995). Pilot-air traffic control communications: It's not (only) what you say, it's how you say it. *Flight Safety Digest, 14*(7), 1–10.

Cushing, S. (1997). Language differences in aviation communication: Problems and solutions. In P. Quigley & P. McElwain (Eds.), *Proceedings of the aviation communication: A multi-cultural forum*.

Dance, F. E. X. (1967). A helical model of communication. In F. E. X. Dance, (ed.), *Human Communication Theory*. New York : Holt, Rinehart and Winston.

Dekker, S. W. A. (2002). Reconstructing human contributions to accidents: The new view on error and performance. *Journal of Safety Research, 33*, 371–385.

Dekker, S. W. A. (2010). Pilots, controllers and mechanics on trial: Cases, concerns and countermeasures. *International Journal of Applied Aviation Studies, 10*(1), 31–50.

Dekker, S. W. A. (2011). Systems thinking 1.0 and systems thinking 2.0: Complexity science and a new conception of "cause". *Aviation in Focus: An international Aeronautical Journal, 2*(2), 21–39.

Dekker, S. W. A. (2012). *Just culture: Balancing safety and accountability* (2nd ed.). Farnham: Ashgate.

DeMatteis, L. (2011). *Site of teaching and linguistic investigation applied in aviation* (Sitio de docencia e investigación lingüística aplicada a la aviación). Online Available HTTP: http://www.lorenadematteis.com.ar/linguistics-and-aviation/aviation-personnel/air-traffic-controllers/?lang=en. Accessed 3 Nov 2011.

DeVito, J. A. (2011). *Human communication the basic course* (12th ed.). Boston: Allyn and Bacon.

Dismukes, R. K., Berman, B. A., & Loukopoulos, L. D. (2007). *The limits of expertise: Rethinking pilot error and the causes of airline accidents*. London: Ashgate Publishing.

Edwards, E. (1988). Introductory overview. In E. Wiener & D. Nagel (Eds.), *Human Factors in Aviation* (pp. 3–25). San Diego, CA: Academic.

Endsley, M. R. (1988). Design and evaluation for Situation Awareness enhancement. In *Proceedings of the Human Factors Society 32nd Annual Meeting* (Vol. 1, pp. 97–101). Santa Monica: Human Factors Society.

Ericsson, K. A. (2002). Toward a procedure for eliciting verbal expression of nonverbal experience without reactivity: Interpreting the verbal overshadowing effect within the theoretical framework for protocol analysis. *Applied Cognitive Psychology, 16*, 981–987.

Garzone, G., Catino, M., Gobo, G., Bait, M., Catenaccio, P., Degano, C., & Rozzi, S. (2010). Towards an integrated model for the understanding of communication failures in aviation accidents: Tenuous identities under pressure. In G. Garzone, & A. James (Eds.), *Discourse, identities and roles in specialized communication* (pp. 208–244). Bern: Peter Lang International Academic Publishers. Online Available HTTP: http://air.unimi.it/bitstream/2434/149610/2/Garzone%20et%20al._2010_offprint.pdf. Accessed 3 Nov 2011.

Gasson, S. (1999). The reality of user-centered design. *Journal of End User Computing, 11*(4), 3–13.

Gasson, S. (2005). The Dynamics of Sense making, Knowledge and Expertise in Collaborative, Boundary-Spanning Design. *Journal of Computer-Mediated Communication* (JCMC), 10(4).

Hayward, B. (1997). *Culture, CRM and aviation safety*. ANZSASI, Asia Pacific Regional Air Safety Seminar, Brisbane.

Helmreich, R. L., & Davies, J. M. (2004). Culture, threat, and error: Lessons from aviation. *Canadian Journal of Anesthesia, 51*(Suppl 1), R1–R4. https://doi.org/10.1007/BF03018331. Online. Available HTTP: http://www.springerlink.com/content/w031115456082773/. Accessed 3 Nov 2011.

Helmreich, R. L., Merritt, A. C., & Wilhelm, J. A. (1999). The evolution of crew resource management training in commercial aviation. *The International Journal of Aviation Psychology, 9*(1), 19–32.

Hofstede, G., & Hofstede, G. J. (2005). *Culture and organizations: Software of the mind* (2nd ed.). New York: McGraw-Hill.

Iani, C., & Wickens, C. D. (2007). Factors affecting task management in aviation. *Human Factors, 49*(1), 16–24.

ICAO (International Civil Aviation Organization). (2000). *Manual of aircraft accident and incident investigation*, 1st Edn/Doc 9756, AN/965. Montréal: ICAO. Online Available HTTP: http://www.aaib.gov.mn/index.php?option=com_phocadownload&view=// 9756_p1_cons_en-4.pdf. Accessed 3 Nov 2011.

ICAO (International Civil Aviation Organization). (2001). *Aircraft accident and incident investigation/Annex 13 to the convention on international civil Aviation* (9th ed.). Montréal: ICAO (Annex 13/Aircraft Accident and Incident Investigation. ICAO Accident/Incident Reporting Manual -Doc 9156). Online. Available HTTP: http://www.emsa.europa.eu/retro/Docs/marine_casualties/annex_13.pdf. Accessed 3 Nov 2013.

Katerinakis, T., Papadopoulos, N., & Zouroudi, D. (2013). Aviation terminology: Literal sense, safety and semantic-oriented Translatology in Greek aviation communication. In P. Kontos, A. Anastasiadi-Symeonidi, M. Katsogiannou, & K. Valeontis (Eds.), *Proceedings of Hellenic language and terminology*. Athens: Hellenic Society for Terminology.

King, C. (1997). Policing language: Linguistic security and the sources of ethnic conflict. A rejoinder. *Security Dialogue, 28*(4), 493–496.

Koch, C. J. (1975). *Small groups in the composition class: A case study of developing linguistic security and written fluency*. Unpublished doctoral dissertation. University of Michigan, Ann Arbor.

Kochan, J. (2009). *HCEP lecture: 'In search of the expert pilot- A research odyssey in human factors'*. Drexel Human Cognition Enhancement Program (HCEP). http://www.drexel.edu/research/humancognition/. Dec 10, 2009. Online Video File at: http://media.irt.drexel.edu/mediasite/Viewer/?peid=8f7355a186f947798bd6bf7db224fefe. Accessed 6 May 2010.

Kvanvig, J. (2006). *The knowability paradox*. New York: Oxford University Press.

Leveson, N. (2012). *Engineering a safer world: Applying systems thinking to safety*. Cambridge, MA: MIT Press.

Leydesdorff, L. (2016). Information, meaning, and intellectual Organization in Networks of inter-human communication. In C. R. Sugimoto (Ed.), *Theories of informetrics: A festschrift in honor of blaise cronin* (pp. 280–303). Berlin/Boston, MA: De Gruyter.

Loukopoulos, L. D., Dismukes, R. K., & Barshi, I. (2001). Cockpit interruptions and distractions: A line observation study. In R. Jensen (Ed.), *Proceedings of the 11th International Symposium on Aviation Psychology*. Columbus: Ohio State University Press.

Loukopoulos, L. D., Dismukes, R. K., & Barshi, I. (2003). Concurrent task demands in the cockpit: Challenges and vulnerabilites in routine flight operations. In *Proceedings of the 12th International Symposium on Aviation Psychology* (pp. 737–742). Dayton: The Wright State University.

Loukopoulos, L. D., Dismukes, R. K., & Barshi, I. (2009). *The multitasking myth: Handling complexity in real-world operations*. Farnham: Ashgate.

MacPherson, M. (1998). *The black box: All new cockpit voice recorder accounts of in-flight accidents*. New York: William Morrow & Company.

Merrin, A. C. (1996). *National culture and work attitudes in commercial aviation: A cross-cultural investigation*. Unpublished doctoral dissertation, The University of Texas at Austin.

Moran Thomas, P., & Anderson, R. J. (1990). The workaday world as a paradigm for CSCW design. In *Proceedings of CSCW '90 Proceedings of the 1990 ACM Conference on Computer-Supported Cooperative Work, Los Angeles, CA, October 07–10, 1990.*

Murphy, M. (1980). Review of aircraft accidents. In G. E. Cooper, M. D. White, & J. K. Lauber (Eds.), *Resource management on the flight deck: Proceedings of a NASA/Industry Workshop* (NASA CP-2120). Moffett Field, CA: NASA-Ames Research Centre.

Nerrière, J.-P. (2004). *Globish the world over – (with videoscript)*. Online Available HTTP: http://globish.com/?page=about_globish&lang=en_utf8. Accessed 3 Nov 2011.

Norman, D. (1988). *The design of every day things*. New York: Bantam Doubleday Dell Publishing Group (Reprint- Originally published: *The psychology of everyday things*. New York: Basic Books Publishers).

Norman, D. A., & Draper, S. W. (Eds.). (1986). *User centered system design: New perspectives on human-computer interaction*. Hillsdale: Lawrence Erlbaum Associates.

Ostrom, E. (2005). *Understanding institutional diversity*. Princeton: Princeton University Press.

Papadimitropoulou, F. (2012). *Establishment of working group on air traffic management (OMEODEK)*. Ministerial decision D4/38654/1084/HCAA.

Person, R. F., Jr. (1999). *Structure and meaning in conversation and literature*. Lanham: University Press of America.

Ragan, P. H. (1994). Subject specialism and general English in aviation ESP. *TESOL Matters, 4*(4), 7.

Roitsch, P. A., Babcock, G. L., & Edmunds, W. W. (1978). *Human factors report on the Tenerife accident*. Washington, DC: Air Line Pilots Association.

Shannon, C., & Weaver, W. (1949). *A mathematical model of communication*. Urbana: University of Illinois Press.

Skybrary. (2010). *Socio-culture (OGHFA BN). Eurocontrol electronic repository/Human factors/Flight safety*. Online Available HTTP: http://www.skybrary.aero/index.php/Socio-Culture_%28OGHFA_BN%29. Stored October 27, 2010, at 22:05. Accessed 3 Nov 2011.

Tompkins, P. K. (1991). Organizational communication and technological risk. In L. Wilkins & P. Patterson (Eds.), *Risky business: Communicating issues of science, risk, and public policy* (pp. 113–129). New York: Greenwood Press.

Ungerer, D. (2004). Simple speech: Improving communication in disaster relief operations. In D. Rainer & J. Kateri (Eds.), *Teaming up: Components of safety under high risk* (pp. 81–92). London: Ashgate.

Vasilescu, I., & Adda-Decker, M. (2007). A cross-language study of acoustic and prosodic characteristics of vocalic hesitation. In A. Esposito, M. Bratanić, E. Keller, & M. Marinaro (Eds.), *Fundamentals of verbal and nonverbal communication and the biometric issue, NATO security through science series E: Human and societal dynamics* (Vol. 18). Washington, DC: IOS Press.

Wegner, D. M. (2002). *The illusion of conscious will*. Cambridge, MA: MIT Press.

Wegner, P. E., Stanley, F., & Jameson, F. (2009). *Life between two deaths, 1989–2001: U.S. culture in the long nineties (post-contemporary interventions)*. Durham: Duke University Press.

Wenger, E., McDermott, R., & Snyder, W. (2002). *Cultivating communities of practice: A guide to managing knowledge*. Cambridge, MA: Harvard Business School Press.

Winograd, T., & Flores, F. (1987). *Understanding computers and cognition*. Boston: Addison-Wesley.

Zexian, Y., & Xuhui, Y. (2010). A revolution in the field of systems thinking – A review of Checkland's system thinking. *Systems Research and Behavioral Science System Research, 27*, 140–155.

Appendix A: The Discussion Sheet of Scenarios

This is the sheet containing the final scenarios used in the third stage of data collection, according to the GT approach used for this study (explained in Chap. 4). The attempts for data collections started with the first-stage interaction of Appendix C; the main theme shifted to the second stage using the questioning route of a fishbone diagram (from conditions to conclusion). In this third and final stage, real-world scenarios were developed and used.

T. Katerinakis, *The Social Construction of Knowledge in Mission-Critical Environments*, Innovation, Technology, and Knowledge Management,
https://doi.org/10.1007/978-3-319-91014-7

Grouping of topics from in-depth interviews. Detailed questions follow:

Grouping of Topics from in-depth interviews

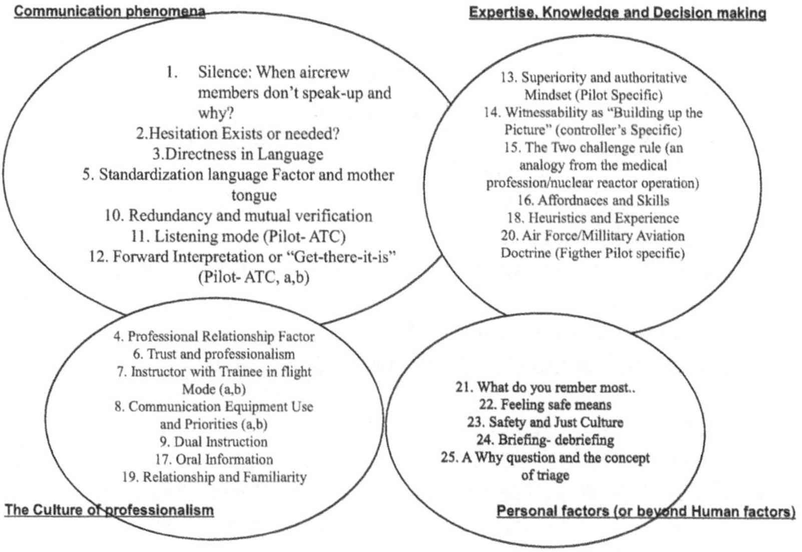

1. **Silence: When Aircrew Members Don't Speak Up and Why (a, b)?**
"Several accidents have shown that crew members' failure to speak up can have devastating consequences. Despite decades of crew resource management (CRM) training, this problem persists and still poses a risk to flight safety. To resolve this issue, we need to better understand why crew members choose silence over speaking up. We explored past speaking up behavior and the reasons for silence in 1,751 crew members, who reported to have remained silent in half of all speaking up episodes they had experienced. Silence was highest for first officers and pursers, followed by flight attendants, and lowest for cap-

tains. Reasons for silence mainly concerned fears of damaging relationships, of punishment, or operational pressures."

Q1a. What is your opinion about the crew members? Did you have such an experience, as a controller expecting a crew member to speak up? What would you do in such a position?

Q1b. Air traffic controllers and "fighter pilot mission supervisors in the tower" are "co-present" in the cockpit. Do you have an experience of wanting to speak but preferred silence? Why? Is there any specific training on how to use/coordinate silence or it is just a matter of rank and authority?

2. **Hesitation Exists or Needed?**

Situation from B737 Captain's Report, http://asrs.arc.nasa.gov/publications/callback/cb_396.html#4

■ *On takeoff roll approaching 80 knots, the Tower Controller called us and said in a very slow, unsure voice, "[Callsign 1...2...3...4...](pause)." He sounded as if he had something to tell us, but did not know what to say. We both noted a tone of concern and hesitation in his voice as if he was still unsure of something at that moment. We were lightweight and had 13,000 feet of runway ahead of us. We had to make an immediate decision.*

The reporter's action: I elected to initiate rejected takeoff procedures. During deceleration the Tower Controller said, "Disregard." The sound of one's voice, the tone, and force, all convey a message. I did not like the message I was receiving and could not gamble that he was trying, but unable, to warn us of something ahead. I would take the same action again.

(From the First Officer's report on the same incident)

I believe the rejected takeoff was the right thing to do. When you get a call from tower at that point in the takeoff roll, the first thing that pops into your mind is "something's wrong." In the few seconds before he finished his thought, we were left to guess what the call was about. We were still relatively slow speed on the roll, so the Captain did what was prudent and safe by rejecting.

Q2. What is your opinion on the ATC hesitation? Did you have such an experience? What would you do if you were the pilot? Is there a way to repair hesitation? Do you think that the first officer should interfere in another way? What is your training on that?

3. **Directness in Language**

In a flight phraseology survey, a pilot commented: "Of course, if I know the ATC person I may bypass the standard phraseology and say "Eh, John this is Tom; what happened, did you hear me? Do you mean it or I have to wait for descent?" If not I may just stay to wait, especially if I do not have a series of corresponding flights in a sequence. I know that safety is the primary concern but we are also professionals who communicate in discussion terms. I feel it as more natural..."

Q3. What is your opinion on the pilot's comment? How direct/informal you think you should be? What is more efficient in your experience? Is it a matter of authority between a pilot and an ATC member?

4. **Professional Relationship Factor**
   A pilot comments: "When I started my career I had a concern, almost fear about the risks that are uncertain when I receive indefinite communication. It took me several hundred hours, I could say after 1,000 flying hours to start felling safe about myself and specific travel routes (when the weather is general terms "stable"). I try to include in my social relations people from ATC, as well as other aviation professionals in order to develop a level of familiarity for their personalities. When I fly abroad, especially overseas I make an effort for an intense briefing about the local controllers, their behaviors and stereotypes in the use of air-speak English and phraseology."

Q4. How do you feel about developing social relationships with pilots flying in your shifts? Is it a trust factor or an obstacle? Do you think that professional dependability is enough, or personal familiarity plays a role? What is your balance on that?

5. **Standardization, Language Factor, and Mother Tongue (a,b)**
   a. Pilot's comment: "For example, when I am in a Greek airport it is really efficient for me and for the safety of my flight to "transcend" SOPs and try to develop a conversational relationship with my controllers (departing and arriving). It is not to violate the handbook but to be laconic and clear when something goes out of the routine. On the other hand I have noticed that ATC people are likely to stay in the letter of ICAO phraseology, they do not seem to feel the liberty or need to deviate."

Q5a. Why do you think that he speaks about a "need" to transcend SOPs? Do you support that "not following the letter of the book" shows that flexibility or that it creates inconsistency? Do you have personal examples to share? What is your experience on ATC personnel? How do you accomplish the "one word – one meaning" phraseology?

   b. "ICAO English may be the spontaneous choice when I follow habitual in-light processes (without too much processing in my mind). I know there is a difference in FAA when I fly in the US as recommended phraseology is not the same. Sometimes, I feel as I am a "recorded voice" in action when I apply the routine SOPs. In contrast when I have to do some more that the standard checklists, I find myself using Greek language and mostly Greek words inside English structures."

Q5b. Do you feel that when you coming from a non-English mother tongue play a role in what you are accomplishing "when communicating"? Is it different in emergencies compared to routine flights?

6. **Trust and Professionalism**

Q6. In the aviation community, most people have faith to each other's professionalism. Do you take professional behavior for granted? Do you play by your role, only? Do you need "interpersonal trust" to the individual (with your crew and the ATC)? How do you develop/communicate that?

7. **Instructor with Trainee in Flight Mode (a, b)**
   Several **reports** have indicated a lack of assertiveness on the trainee's part and a failure to challenge the instructor even when the trainee believed the instruc-

tion was wrong. The following study report excerpt exemplifies how confusing and vague communication by both instructor and trainee can result in a safety incident:

a. Instructor said… "Uh, you can have control if you, uh, want it." I probably replied "OK" rather than the usual "I have control." I began to pull the nose up slowly when I thought I felt my instructor push forward on the wheel [and] relaxed…Nosewheel touched down first and we bounced…Fortunately we walked away…with an undamaged aircraft. "Wishy washy" coms played a major role in this (ASRS Rec. #240165).

Q7a. Have you been to such a position? What is your opinion (in the position of an instructor or trainee)?

b. It is a common technique of flight instructors to allow the trainee to make mistakes in an attempt to develop independent actions and observe the trainee's level of awareness. However, especially during IFR operations, or when compliance with an ATC directive is doubtful, corrective verbal comments by the instructor have a significant impact on flight safety.

Q7b. What is your opinion (as instructor or trainee)? How do you develop your communication to repair such cases?

8. **Communication Equipment Use and Priorities (a, b)**

An instructor's testimony:

a. "We had started flying using headsets, with the radios being monitored through the headsets. After the first landing the student stated he would prefer to continue without the headsets as he didn't feel comfortable wearing them. I said OK. We got involved in doing touch and goes (5 of them) and I failed to notice that we had not heard from Tower during this time. When I did notice that the speaker button was not in the proper position, I made contact with the Tower. They (Tower) terminated the flight and I was instructed to call the Tower." *(ASRS Record #290210)*

Q8a. How confident are you to fly without listening to the tower? What would you do in such a position?

b. An instructor's perception of task priority may have been distorted by the desire to critique the student and resulted to wrong-runway takeoff:

"We took off on [runway] 24 instead of 30, as the Tower subsequently informed us. As I reviewed the event later, with my student and in my own mind, I realized how I may have added to the uncertainty. I was busy pointing out airport markings and critiquing the flight to this point. The priority should have been communications with the Tower and standard procedure." *(ASRS Record #137322)*

Q8b. How do you stay focused on the instruction flight? Did you have such an experience? How do you keep up with priorities during flight?

9. **Dual Instruction**

Reports suggest that the operational context in which dual instruction often occurs, specifically, the *simultaneous occurrence of internal verbal or external radio communications with aircraft maneuvers and demonstrations*, may be an incident-generating factor. But it is also helpful for instructors to remember that every word counts – as well as the timing of training-related critiques. For example, it is more effective for an instructor to say "*turn left 90 degrees*," than to ask, "*where are you going?*" as the aircraft enters controlled airspace without a required clearance. Studies suggest the need to improve the clarity, economy, and judgment of priority of verbal communications in dual training, especially for flight instructors. Trainees need to be able to express doubt or uncertainty and also to admit mistakes.

Q9. How do you accomplish such tasks? Could you describe briefing and debriefing techniques to do so, or you prefer to do it during the flight? Do you feel comfortable with multitasking?

10. **Redundancy as Mutual Verification**

In several transcripts, a pilot misreads the numbers in a clearance message and repeats back the erroneous units for controller confirmation. A controller did not hear – or did not listen to – the incorrect readback. The airman accepted lack of response as silent confirmation that the readback was correct. Inadequate "*Roger*" *or* "*okay*" *or* "*so long*" types of pilot acknowledgments for clearances that precluded any controller double check of the completed exchange.

Q10a. Do you think that readback-hearback interactions create more fatigue and are monotonous, or you take them as necessary? Is it plausible to make such inferences in flight or it is just a matter of experience?

Q10b. An experienced flight instructor commented in a report: "misread, turned out, guessed, hearing one number and saying back another, hearing what I expected to hear, forgetting to change the altitude reminder to a new altitude, even ascertained by an informal consensus vote in the cockpit," which are common occurrences in flights.

What are the "rules of thumb" to repair or prevent them? Is it part of the unexpected to deal with?

11. **Listening Mode (Pilot-ATC)**

A pilot complained in a collective report: "*while it is possible that we may misunderstand the controller, nevertheless we were relying upon the controller to correct any mistakes in the readback. My impression is that controllers are not in the listening mode. As soon as they issue a clearance, they start talking to other aircraft and pay no attention to the readbacks!*"

Q11. Did it happen to you? Are there any problems in prioritizing "listening" vs "speaking" and "readback and hearback" more between different flights? Are you involved in any type of training for that?

12. **Forward Interpretation or "Get-There-It-Is" (Pilot-ATC, a, b)**

a. A controller in a collective report: "*In my experience, flight crews are prone to interpret an advisory of traffic at some altitude as clearance to that altitude. If flight crews plan to request a specific altitude they tend to translate the next clearance to that altitude.*"

Q12a. Do you think that pilots tend to be furious and take the next step for granted? Is it a phenomenon of "get-there-itis"; i.e., I want to arrive rather sooner than later (for the next position and for the final destination)?

b. If we are aware of this "get-there-it-is," we can break the domino effect, an effect that takes place in every accident related to pilot error, one bad decision, leading to another bad decision, leading to an even worse decision, and inevitably ending with severe consequences.

Q12b. What is the type of training/experience that helps you avoid such a domino effect? How do you communicate a change of a decision/repair a bad decision? What is the instruction and the practice to repair such cases?

13. **Superiority and Authoritative Mindset (Pilot Specific)**

Being a Captain: *The flight officer had a background of military fighter aircraft and is used to being his own commander and taking the initiative. I was hesitant to give him hard orders, so I let him make several moves before I was ready for them. Later, I had a long conversation with him as to who was the "Captain."*

Being the co-pilot: *The captain was so sure of himself that it would have been presumptuous for me to ask him to verify the information. The next time, I will find a way to do it regardless of the Captain's attitude.*

Q13. Did you have an experience of conflicting roles? Is there a mindset of authority or rank? Is it different between men and women in those positions? What is your perception and what would you do?

14. **Witnessability as "Building Up the Picture" (Controllers' Specific)**

Air traffic control can be viewed as collaborative activity. Controlling air traffic is managing events in real time rather than following the kind of plan which determines in advance what it is the controller should do next. An important part of this is "looking about 5 to 10 minutes ahead all the time….sometimes a little further than that," in order to get an idea of the actions needed to be taken "now." As one controller described the process at some length, *the kinds of check you do is check your information which is your strips which will show you whether an aircraft is in there or not, and that means then that you don't have to look all over the radar… it would be an impossible job to sit down and look at the radar and look at all the different blips and try to avoid them by putting the aircraft into blank spaces on the radar, so you have got to have this information to tell you what traffic is coming into and out of the sector.*

Q14. What is the communication used during this type of scheduling? Are there any time gaps during this communication? Do you use any visual or other practical aids like "paper strips"? How do you hold/remember that type of information?

15. **The Two-Challenge Rule (An Analogy from the Medical Profession/ Nuclear Reactor Operation)**

The two-challenge rule allows one crew member to *automatically assume the duties of another crew member who fails to respond to two consecutive challenges.* For example, the pilot-on-the-controls becomes fixated, confused, and task overloaded or otherwise allows the aircraft to enter an unsafe position or attitude. The pilot-not-on-the-controls first asks the pilot-on-the-controls if he/she is aware of the aircraft position or attitude. If the pilot-on-the-controls does not acknowledge this challenge, the pilot-not-on-the-controls issues a second challenge. If the pilot-on-the-controls fails to acknowledge the second challenge, the pilot-not-on-the-controls assumes control of the aircraft.

Q15. Do you think that you could apply that rule? Can you think of any examples?

For ATC: Is that happening between controllers' disagreements (between area approach and airport tower or other sectors of air traffic control)?

16. **Affordances and Skills**

Flight operation and air traffic control is considered a mission critical environment because of human safety involved, high stakes in resources, time constraints, and the mission involved in military aviation.

Q16. What helps you operate better under pressure when time is critical? What is the level of training or experience that facilitates coping with overlapping communication and decision-making under pressure and limited time?

17. **Oral Information**

Experienced practitioners in their daily practice, rather than totally relying on standard operating procedures, make use of their own tacit experiences when faced with decisions to make.

Q17. Do you happen to obtain information/knowledge/expertise about your work orally or from manuals and the books? Are there any cases that you would like to share?

18. **Heuristics and Experience**

It is widely accepted that mastering most complex human endeavors (from chess playing to medical intensive care) requires a 10-year experience. What is the experience that you would describe as necessary and adequate in your sector? Does it mean to be able to interpret SOPs and rules of engagement (RoE) to solve problems and make judgments quickly and efficiently, but they are also prone to errors?

Q18. How do you communicate your situated knowledge when changing shifts? Do you like to write short notes as reminders? What are the conditions of affecting performance to hold?

19. **Relationship and Familiarity**

As common theme in discussions with pilots and air traffic controllers, a typical conclusion is that "having gone to the flight school/air force academy together/being in the same crew many times, being in the same shift many times/being in the same dyad or quadruplet or octave of fighter planes together creates a type of closeness and proximity that facilitates communication and mission completion."

Q19. How do you communicate with your "buddies" from those cases? Does it help you bypass standardized patterns? Can you think of any examples from verbal or nonverbal communication? What happens when crews and shifts change? Did you have any relevant experiences in international missions/flights?

20. **Air Force/Military Aviation Doctrine (Fighter Pilot Specific)**
In the case of fighter pilots, many accounts connect the political and tactical training culture with effectiveness in mission completion and operating under pressure. Furthermore, *when training is focused on defensive missions and the broader political surroundings promote a defensive doctrine, fighter pilots seem to have an esoteric motivation that goes beyond their operational mandate.*

Q20. Is that happening in your experience when flying in NATO/international missions compared to national missions? Does it affect the way you operate/communicate during flights? Is it the mission or what you feel about it? Do you connect "mother tongue" with "mother land" when you communicate? How do you manage the multiplicity of communication from Air Force command, traffic control, mission chief, etc. that accumulate?

21. **To Remember…**
Is there any flight/example/case that you always remember, so far? Why?

Q21. Do you fly with a "plan B" in mind or in your briefing?

ASRS #409: "I have flown many years and I am very comfortable flying VFR and IFR, even VFR when the ceiling is low as long as the visibility is as good as it was this day. However, I let my comfort level lull me into departing without a viable Plan A and no Plan B."

22. **Feeling Safe**

Q22. In a more general idea, what makes you feel safe in communication (from one shift to the other in ATC and with incoming pilots)? What type of briefing do you think is efficient?

23. **Safety and Just Culture**

Q23. What is the meaning/your understanding/perception of safety and just culture for you?

24. **Briefing-Debriefing**

Q24. What is the importance of "briefing-debriefing" in your functions? Does it happen to complement your briefing during a flight?

25. **A Why Question**

Q25. Why did you choose to come to the ATC profession? Why did you become a pilot? What is/was your vision/goal when you started?

Note: Write any comment you may have for the questions or the cases of this "interview."

Thank you for your cooperation
Theodore Katerinakis ■

Appendix B: Ishikawa Diagram and Questioning Route

This is the diagram containing the constituent questions in the second stage of data collection, according to the GT approach used for this study (explained in Chap. 4). The second and final shift on the main topic occurred in the transition to the scenarios (in Appendix A). Fishbone diagrams (also called Ishikawa diagrams or cause-and-effect diagrams) are visual tools that show the path for contributing factors to a certain event (area of interest or problem). The main arrow or trunk represents the main goal, and primary factors are represented as sub-arrows or branches. Stems, leaves, and so on are added as secondary factors and tertiary factors, respectively.

The Ishikawa Diagram

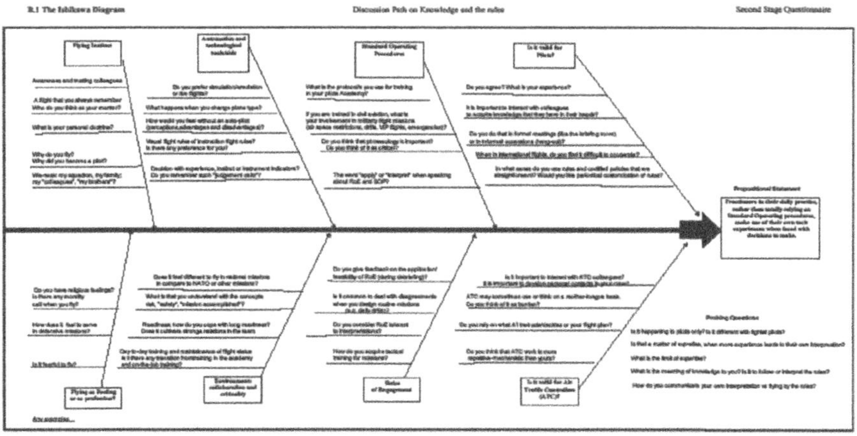

© Springer International Publishing AG, part of Springer Nature 2019
T. Katerinakis, *The Social Construction of Knowledge in Mission-Critical Environments*, Innovation, Technology, and Knowledge Management,
https://doi.org/10.1007/978-3-319-91014-7

Questioning Route of the Ishikawa Diagram

Theoretical theme (in the "fish head"): Practitioners in their daily practice, rather than totally relying on standard operating procedures, make use of their own tacit experiences when faced with decisions to make.

Is it valid for pilots? In order to draw the frame for pilots as highly skilled practitioners in an environment of mediated communication and explain communicative activities that they commit.

Do you agree? What is your experience?
It is important to interact with colleagues to acquire knowledge that they have in their heads?
Do you do that in formal meetings (like the briefing room or in informal occasions (hang-out in the squadron)?
When stationed in other bases, do you find it difficult to cooperate?
In what sense do you use rules and codified policies that are straightforward? Would you like periodical customization of rules?

Is it valid for air traffic controllers? In order to show how pilots (senders) perceive or understand the position of ATC (receivers) in their communication interaction. This sequence could be used to validate questions.

Is it important to interact with ATC colleagues?
It is important to develop personal contacts in your squadron?
ATC in your missions has an additional superior flight officer to conduct your coordinated missions. Do you feel that necessary for ATC?
Do you rely on what ATC suggests/decides or you prefer to follow your mission briefing?
Do you think that ATC work is more repetitive-mechanistic than yours?
Does it happen to interpret ATC guidelines on your own? Do you need to adjust your flight plan/route with your own initiative?

Standard operating procedures: to point out what are the standards in their operation, how they acquire communication competence and data rules, as well as how they perceive their "liberty to interpret"

What is the protocol(s) you use for training in your Pilots Academy?
If you are trained in civil aviation, what is your involvement in military flight missions (air space restrictions, drills, VIP flights, emergencies)?
Please describe typical attributes/examples of phraseology used in flights
Do you think of it as critical?
The word "apply" or "interpret" when speaking about RoE and SOP?

Rules of engagement: to explain what makes it different in the rules of specific mission (military aviation) compared to standard processes

How do you acquire tactical training for missions?
Is it common to deal with disagreements when you design routine missions?
Please list some examples of RoE. What is your opinion about those rules? How different are they from SOPs? Is there room for your own decision or tolerance?
Do you give feedback on the application/feasibility of RoE (during debriefing)?

Automation and technological tools/aids: to investigate trust in automation, interactions with autopilot, and reliance on critical thinking/expertise on input data

Do you prefer simulation/emulation or live flights? How often do you use the auto pilot?
How would you feel to fly a drone? (perceptions, advantages, and disadvantages)
Visual flight rules of instruction flight rules? Is there any preference for you?
Decision with experience, instinct, or instrument indicators? Do you remember such "judgment calls"?

Environment: collaboration and criticality, to investigate the context of the flight communication model in different stages of an aviation career

Does it feel different to fly in national missions compared to NATO or other missions?
What is that you understand with the concepts risk, "safety," and "mission accomplished"?
Readiness: How do you cope with long readiness? Does it cultivate strong relations in the team
Day-to-day training and maintenance of flight status Is there any transition from training in the academy and on-the-job training?

Flying instinct: to investigate deeper characteristics that define the pilot's personality and affect their communication and team spirit

Awareness and trusting colleagues
A flight that you always remember
Who do you think as your mentor?
What is your personal doctrine?
Why do you fly? Why did you become a pilot?
We-ness: my squadron, my family, my "colleagues," "my brothers"?

Flying as feeling or a profession: emotional profile of a pilot in regard to implicit communication

Is it fearful to fly?
How does it feel to serve in defensive missions?
Do you have religious feelings for what you do?
Is there any morality call when you fly?

Appendix C: "EGO-ALTER" Interaction with an Adjustable Questionnaire

This is a transcript of an interaction used in the first stage of data collection, according to the GT approach used for this study (explained in Chap. 4). The theme of the discussion was experience, explicit knowledge, and creative deviance from the rules. The first shift on the main topic occurred in the transition to the Ishikawa diagram and questioning route (in Appendix B).

Ego: *Hello, Captain A and thank you for participating in this study. I want to thank you for taking the time to meet (e-meet)/deal with my questions, and I would like to talk about your flight experiences and specifically the use of standard operating procedures and the extent to which individual initiative is necessary in aviation decision-making skills.*

All responses are confidential, and any information included in the study does not identify you as a respondent.

Alter: *I understand as I was briefed it with Colonel K, when you first visited the base. How do you know him?*

Ego: *I was serving my duty in 115 CW during the Imia crisis and I was a sub-standard during the week of COC missions.*

Alter: *Nice, I was in the Academy at the time…*

Ego: *Coming from that experience I wanted to accent the importance and role of HAF pilots as highly skilled, risk-taking practitioners and decision-makers. Furthermore, I would like to explore the limitations, tolerance, or creativity involved when following SOPs and RoEs. Lastly, to explore the role of communication in an efficient implementation of a flight mission.*

Alter: *these are kind of esoteric questions that we keep in the back of my head, since when I am about to fly I never know what my experience would be.*

Ego: *I know, I know …and I thought to use some real cases to have a base for this discussion.*

Alter: *(nonverbally knocking head to go on)*

Ego: *Here is a situation reported in NASA's Aviation Safety Reporting System:*

T. Katerinakis, *The Social Construction of Knowledge in Mission-Critical Environments*, Innovation, Technology, and Knowledge Management,
https://doi.org/10.1007/978-3-319-91014-7

A real case of a *B737 captain's report*: On takeoff roll approaching 80 knots, the Tower Controller called us and said in a very slow, unsure voice, "[Callsign 1…2…3](pause)." He sounded as if he had something to tell us, but did not know what to say. We both noted a tone of concern and hesitation in his voice as if he was still unsure of something at that moment. We were lightweight and had 13,000 feet of runway ahead of us. We had to make an immediate decision.

Ego: *What is going on here, in your opinion?*

Alter: *Oh… that is not so uncommon situation [laughing]. I am not sure why they report it.*

Ego: *What do you mean? This is the ASRS anonymous system of reporting. What are the rules for that?*

Alter: *So, when we are in the flight school we share training ours with ATC so that we understand their work. They have multiple stimuli in their screens and visual scope and each communication unit is a separate case.*

Ego: *[interrupting] Is or should be? Do they have to multitask?*

Alter: *Of course, they do but they have resources to isolate frequencies, call signs, one flight after another and also to categorize intra-flight communications (from pilot to copilot, etc.).*

Ego: *I see, but let's go back to the rules. What are the SOPs applied in this situation.*

Alter: *I have to say that in such a discussion I feel like a doctor who evaluates another doctor's patient, without being there. But, this is straightforward to say that I should ask back to ATC, before reporting back on the command, "what happened? Repeat Tower?"*

Ego: *Do you just repeat, as we would do in routine conversation? What about the tone of the voice and hesitation?*

Alter: *If it is not incidental (as it could be), it could be a warning sign for an uncomfortable situation, a problematic pendency or something that has to do with the actual flight, or the status of the runway, or the taxiway.*

Ego: *So how do you evaluate that on site? Do you just follow the book or try to relate more with the ATC person?*

Alter: *It is quite different is you are airborne or you are preparing for clearance. It seems that this pilot is still on the ground, isn't he?*

Ego: *Correct, he was in the ground. But, how do you decide?*

Alter: *At first, when on the ground, if I have a co-pilot I consult him for his opinion as it is common for us to challenge the ATC when in small airports. If you are airborne that is a different story.*

Ego: *Is it so objective to tell, about a busy or a small airport or how you relate with ATC?*

Alter: *Of course, if I know the ATC person I may bypass the standard phraseology to say "Eh, John what happened? Why do you hesitate?" If not I may just stay to wait, especially if I do not have a series of corresponding flights in a sequence. I know that safety is the primary concern but we are also employees…*

Ego: *So, do you think that this is judgment call? What are your criteria? Are they different from the book?*

Alter: *when I started my career I had a concern, almost fear about the risks that are uncertain when I receive indefinite communication. It took me several hundred hours, I could say after 1,000 flying hours to start felling safe about myself and specific travel routes (when the weather is general terms "stable"). I try to include in my social relations people from ATC, as well as other aviation professionals in order to develop a level of familiarity for their personalities. When I fly abroad, especially overseas I make an effort for an intense briefing about the local controllers, their behaviors, and stereotypes in the use of air-speak English and phraseology.*

Ego: *Do you really need to go so deep when you prepare yourself for a flight? [with naiveness] I thought ICAO phraseology and standardized training guarantee uniformity...*

Alter: *we are not in an ideal world, and we are still humans. Pilots and air-traffic controllers are social animals with training and experience but with preferences and prejudices, also. For example, when I am in Greek airport, it is really efficient for me and for the safety of my flight to "transcend" SOPs and try to develop a conversational relationship with my controllers (departing and arriving). It is not to violate the book but to be laconic and clear when something goes out of the routine. On the other hand I have noticed that ATC people are likely to stay in the letter of ICAO phraseology, they do not seem to feel the liberty or need to deviate.*

Ego: *Why do you speak about a need? Do you support that "not following the letter of the book" you are more flexible?*

Alter: *No, it is not exactly that. I say that rules may come to an end in a situation of escalating crisis or even when you have multiple warning signals that have to be communicated. It is ironic, sometimes, that after all this complexity that we experience and operate in a flight cockpit and plane "piloting" is implemented only when we communicate our acts and we communicate to act.*

Ego: *So, do you feel that coming from a non-English mother tongue plays a role in what you are accomplishing "when communicating"?*

Alter: *I would say yes and no; my point is that language (and mainly phraseology) may affect my understanding and the messages I convey but my Greek experience seems quite embedded in my language skills. Of course, I have gone over various ICAO certifications, but still a lot of times it is spontaneous to think in terms of "Greek reality and mentality" since I need to fly to diverse small airspaces for domestic flights and accomplish some type of rapport with different controllers. ICAO English may be the spontaneous choice when I follow habitual in-light processes (without too much processing). Sometimes, I feel as I am a "recording" in action when I apply the routine SOPs. In contrast when I have to do some more that the standard I find myself using Greek language and mostly Greek words inside English structures.*

Ego: *That means that in this case of hesitation, would go to a direct Greek comment/inquiry to use?*

Alter: *It is quite likely, especially if I was flying when time is critical and decision-making space is constraint.*

Ego: *But in this case, it is how you hear the other talk and not so what does he say?*

Alter: *We are instructed and trained to be more rigorous and economical when we communicate; every message has to have a clear start and end. But, since I receive human voice in my headphones, it is implicit communication with an element with a bit of trust.*

Ego: *In the aviation community I have seen that most people have faith to each other professionalism. Why do you need trust?*

Alter: *I agree that you have faith to your partners – that how I feel about all participants in a flight mission – because that is what we do. I use technology, I use tools, I use communication in systems that are dependable, and I have faith… but with ATC I need something more, I need to establish trustworthiness in what the other person does and trust in what he/she communicates (sometimes I have experienced differences when speaking in a female ATC in compare to man but it is another story). When I need guidance, when I need confirmation, when I need a solution I cannot have second thoughts. I need to feel that there is no question in what ATC is saying. Flying is not an individual accomplishment; it needs collective work on board the plane but also a relationship continuum with the controllers. That is why when I know their first name, I use it, although it is not recommended. This concept is more important to be accomplished in routine iterations of flights, as you may only need to make use of it when an incident is unfolding, an event may be progressing, or an emergency is diagnosed. It is part of the idea of being proactive, to know your resources and especially the human resources you can rely on.*

Ego: *that's really useful captain. Let's see what that pilot and copilot reported in action:*

The reporter's action: *I elected to initiate rejected takeoff procedures. During deceleration the Tower Controller said, "Disregard." The sound of one's voice, the tone, and force, all convey a message. I did not like the message I was receiving and could not gamble that he was trying, but unable, to warn us of something ahead. I would take the same action again.*
From the First Officer's report on the same incident: *I believe the rejected takeoff was the right thing to do. When you get a call from Tower at that point in the takeoff roll, the first thing that pops into your mind is "something's wrong." In the few seconds before he finished his thought, we were left to guess what the call was about. We were still relatively slow speed on the roll, so the Captain did what was prudent and safe by rejecting.*

Ego (continues):	*What is your opinion?*
Alter:	*Better safe than sorry is a canon. The pilot perceived a level of uncertainty. Since they were on the ground there is no safety issue at stake, so a delay is a prudent choice and action, as the first officer suggests. I am not sure whether I would do the same thing or not. If I was in a Greek airport, I would try to resolve this "uncertainty hesitation" with a more informal inquiry, i.e., to ask back and clarify the sources of this hesitation. Still, in EU Airspace as governed by Eurocontrol delays may have domino effects in consequent routes (mainly in busy airports). That means that if you lose your time stamp to depart, you may need to wait for the next cycle of allocated expected departure times that may be quite disturbing if you have consequent flights.*
Ego:	*That a really thoughtful comment from my theoretical point of view also. Would you like to share any example of yours in a similar case?*
Alter:	*I have not experienced that so far, as I try to be persistent with ATC when there is an uncertainty. But, I always have in mind the tragic collision in Tenerife Airport on 1977 when the crash took place on the ground!!!*
Ego:	*Is there any flight that you always remember?*
Alter:	*That is the flight that stays in my mind to remain proactive. I also keep aside to complement, my spiritual beliefs although I cannot articulate it with facts but I reserve it as a contributing factor for a peace of mind.*
Ego:	*In a more generic idea, what makes you feel safe in communication?*
Alter:	*Feeling safe in communication means to feel during the flight. Flight briefing in carrier's headquarters is very important, meeting my crew members is really useful, and being able to start conversation with informal tips with ATC. Also, I need to follow the inflight takeoff checklist and study the flight log for potential entries and problem resolution. Thus, I need to go by the book with SOPs in order to be able to have choices in my decision-making. Sometimes, eliminating alternatives may be more important than having more, especially when time is counting down. Lastly, when I have the option (or tolerance), to speak briefly in my mother tongue, that is another assurance in my mind.*
Ego:	*Let's recap some key points; safety is a doctrine for you. Communication is a key tool to accomplish a flight. I daily practice, rather than totally relying on standard operating procedures, you make use of your own experiences when faced with decisions to make. To communicate in Greek, maybe also*

necessary in a limited form , especially when your experience allows you to go beyond self-reliance in automation.

Alter: *I wouldn't recap it that way but this is a fair statement to conclude.*

Ego: *Is there anything, you would like to add?*

Alter: *I would like to point out that is a very important topic for the aviation sector and your collection would be a very good start to initiate an ASRS system to be constructive in everyday life of aviators.*

Ego: *Thank you very much, for your time and I will keep you posted.* ■

Appendix D: ICAO-FAA Basic English Phraseology Differences

The use of aviation English challenges cockpit communication in international flights. Differences in phraseology generate a division by a common language.

Differences

FAA	ICAO
Point	Decimal
Ramp	Apron
Airport	Aerodrome
How do you hear me?	How do you read?
Affirmative	Affirm
Attention all acft	All stations
Taxi back	Back track
Closed Traffic or Pattern	Circuit
Traffic no factor	Clear of traffic
Verify	Confirm
Continue	Continue Approach
Continue heading	Fly present heading
Center (ARTCC)	Control (Area Control Centre)
Clearance Delivery	Clearance
Descend at pilot's discretion	Descend when ready
For spacing	For separation
DF Station	Homer
Radar Contact	Identified
Radio (FSS)	Information
Enter downwind	Join downwind
Taxi into position and hold	Line up and Hold
Maintain VFR	Maintain own separation and VMC
Remain in closed traffic	Make another circuit
Make one three sixty	Make an orbit
Resume normal speed	No ATC speed restriction
Radar Contact	Observed position
Leaving	Passing
PAR	Precision
Loud and clear	Read you five
Squawk altitude	Squawk charlie
Hold	Stop
Departing	Taking off
Course	Track
Exit	Vacate

T. Katerinakis, *The Social Construction of Knowledge in Mission-Critical Environments*, Innovation, Technology, and Knowledge Management,
https://doi.org/10.1007/978-3-319-91014-7

Appendix E: Naming Conventions and Profiles

Discussions were recorded with the naming convention of airport names to preserve anonymity (46 airports are documented and supervised by the Hellenic Civil Aviation Authority and 12 combat wings, without using all of them; in number #25 "Stan Dardman" is a conventional name). That was the second level of coding the names, as all names were initially codified with a conventional acronym: *Aviation Informant Professional <number>.*

Eleftherios Venizelos AIP1, Fighter Pilot	13.	Megas Alexandros – AIP13, Accident Investigator
Ioannis Daskalogiannis AIP2, Fighter Pilot	14.	Ikaria Ikaros AIP14, ATC
Thessalonikis Makedonias AIP3, Fighter Pilot	15.	Oia Santorinis AI1P5, ATC
Nikos Kazantzakis AIP4, Fighter Pilot	16.	Vasileios Pyrros AIP16, ATC
Ioannis Kapodistrias AIP5, Fighter Pilot	17.	Omiros Chios AIP17, ATC
Ippokratis Koos AIP6, Helicopter pilot	18.	Filippos Kozanis AIP18, ATC
Democritos Alexandroupolis AIP7, Fighter Pilot	19.	Aristotelis Kastorias AIP19, ATC
Odysseas Elytis AIP8, Civil Aviation pilot	20.	Alexandros Onassis AIP20, Flight Engineer
Ifaistos Limnios AIP9, Civil Aviation pilot	21.	Dimitrios Vikelas AIP21, ATC
Diagoras Rodios AIP10, Civil Aviation Pilot	22.	Vassilis Konstantakopoulos AIP22, ATC
Dionysios Solomos AIP11, ATC	23.	Neos Aghialos AIP23, Fighter Pilot
Aristarchos Samios AIP12, ICAO delegate	24.	Palios Ellinikos AIP24, Civil Aviation pilot
	25.	Stan Dardman AIP25, Civil Aviation Pilot

T. Katerinakis, *The Social Construction of Knowledge in Mission-Critical Environments*, Innovation, Technology, and Knowledge Management,
https://doi.org/10.1007/978-3-319-91014-7

Profiles and Background of the Responders

Eleftherios Venizelos	HAF fighter pilot with more than 20 years of flight mission experience, a second University degree, and NATO AWACKS Crew Supervisor. He graduated from the Air Force Academy with a less adventurous ambition in comparison with his pilot colleagues; instead of being obsessed with the latest technological breakthrough in aircraft hardware, he was inspired by the communal living and close ties formulated in the flight school and wanted to dedicate his learning on how dyads of pilots or group of pilot classmates would become competent and reliable professionals. After spending 10 years in A-7 squadrons, he graduated with a second university degree in decision sciences. He spent several years in HAF General Staff and other 8 years in NATO Reconnaissance Squadrons representing Greece. His missions were designed to aviate, as Chief of the Strategic Design Unit on board, with a team of international crew members in AWACS aircrafts, before returning to serve in the HAF General Staff again, in a new branch. He had to undertake diverse missions, work in multicultural and multilingual military environments, and developed a strong sense of collaborative work in several thousand flying hours. He has active experience in national fighter missions in HAF, in international reconnaissance missions for NATO where he needed to take off and land in most of the 17 countries which consists the operational branch of NATO. He believes that Air Force works because of the groups of people that are calibrated in sophisticated teamwork after intense training
Ioannis Daskalogiannis	HAF fighter pilot with more than 25 years of experience in elite units, two postgraduate degrees, and head of HAF Academy curriculum development. Strong background in aviation safety and international fighter missions. A HAF fighter pilot at the rank of Brigadier General with 20 years of operational experience in 4,000 flying hours on board of T-2s, A-7s, and Mirage 2000–5, which are considered the elite unit of HAF. He holds postgraduate degrees from the US military schools and has served in the Hellenic Air Force Academy where he redesigned several parts of the curriculum. His involvement on tactical and operational training for different generation of pilots oriented his focus on mentoring, tactical design, and international allied missions. His bonding with the younger officers is exceptional; he has spent 3 years on consecutive shifts every second day in his base and formulated a tight bond with the younger pilots who were in the tactical training phase
Thessalonikis Makedonias	Brigadier General HAF, in commanding positions of a NATO major installation, with 30 years of flight experience. Served in commanding positions of fighter squadrons and pilot training squadrons with A-7s. He has substantial experience in international culture and mission coordination
Nikos Kazantzakis	Brigadier General (HAF) just retired due to health reasons, with commanding experience in fighter squadrons and pilot training squadrons (A-7s). Strong interest in pilot safety, with research work on the "supervisor's error" in advanced mission design and training
Ioannis Kapodistrias	Brigadier General (HAF), with very extensive commanding and managerial experience (in NATO's Allied Joint Force Command Naples) and HAF combat wings (in A-7s). With a second degree in political science and a strong awareness on public policy and personnel management

Ippokratis Koos	Colonel of Hellenic Army Aviation, with commanding experience at a battalion level and specialization in tactical design. A helicopter pilot known for effective crew coordination. He is a key partner in NATO simulation game design. He is specialized in Chinook helicopters and serves with close cooperation with Apache attack helicopters. After spending 15 years in conventional Army operations out of the Hellenic Army Military Academy, he has experienced a chain of command established without flexibility and a tradition of hierarchical organization and ethos that goes back to ancient historical warfare, in the epic works of Homer, Thucydides, and Xenophon
Dimokritos Alexandroupolis	Major General (HAF), with commanding position in three combat wings, the Hellenic Tactical Command, and the HAF General Staff. He has an expertise in allied missions and international drills and participates in the design of NATO simulation war games. He has spent most of his flying hours in F-16s
Odysseas Elytis	Civil Aviation pilot with 25 years of experience in passenger planes in one of the two largest airliners in Greece. His flight experience includes routes all over Europe, the Middle East, and the Arab States. Most of his flight hours are in B-737 aircrafts
Ifaistos Limnios	Civil Aviation pilot with 20 years of experience in passenger planes and more than 10 years in VIP planes. In the past 10 years, he is flying in European countries and Russia. He has a personal experience from major aviation accidents, accident investigation, and the respective litigation
Diagoras Rodios	Civil aviation pilot with a strong experience in various aviation topics (radio communication, search and rescue communications equipment in military and commercial environment, and air traffic control). His expertise and passion is aviation history and has experimented with civilian unmanned vehicles. He has an authentic drive for aerial inquiries and considers aviation an ultimate goal in life. He started from after high school in the vocational training school for technicians in aviation engineering and continued with degree in radiotelephony and radio communications (commercial, in HAF and Hellenic Coast Guard). In a period of 15 years, he obtained flight degrees from schools in the USA, France, and Greece, as well as flight simulator instructor degree, while working as a commercial pilot in Greece. He coordinated the single-day transition from one airport to the other Athens Airport to the new "Eleftherios Venizelos" Airport and supervised the air traffic control in ten peripheral airports of the Aegean islands for quality assurance
Dionysios Solomos	A top executive in HCAA, with a specialization in environmental engineering in aerodrome installations and a strong managerial and supervisory experience
Aristarchos Samios	A top executive in the Supervisory Council for Air Navigation, with an expertise in aviation economics and public infrastructure design. He has spent 5 years as an administrative member of the HAAISB and as a national delegate to ICAO
Megas Alexandros	An experienced accident investigator with more than 140 accidents investigated, after serving in HAF and employed in civil aviation, with thousands of flying hours in three generations of aircrafts. He graduated from Air Force Academies in Greece, USA, and Sweden and specialized in accident investigation in three different countries. He has a strong experience in international aviation coordination in Europe, USA, and Asia and has provided important safety recommendations in the industry. His safety/technical submissions and accident investigation reports are significant inputs for key players in the industry (e.g., Boeing, Dassault, Honeywell, among others). Served in ICAO delegations and task forces for more than 15 years

(continued)

(continued)

Ikaria Ikaros	The most experienced air traffic controller in this study, with active participation in SPOA and EASA task forces. One of the first women in air traffic control in Greece, who is also married to an air traffic controller. She has an expertise in aviation regulatory framework in Europe and relevant public policies. After having a 30-year experience in the sector, she is one of the first few women in the profession in Greece, coming from a background in law. She served as delegation member in the European negotiations for the Single European Sky initiative and is still an instructor in the Civil Aviation School. She is quite sensitive in the use of airspeak English and airspeak Greek
Oia Santorinis	A female young air traffic controller, with a degree in international relations and fluent in three languages. She has an active participation in aviation terminological research. A young controller with a different profile, both in the admission point as well as in her current work. She is a woman again but entering a professional world with a rapid growing percentage of 30% and with the first woman Head of the HCAA to assume position in the current decade. She is a woman again in the increasing number of women in the professional world of aviation. The first woman Head of the HCAA assumed position in the current decade. The percentage of women in the sector is 30% and growing. Her career aim is to continue to accident investigation and to fill as many positions as possible in peripheral airports before assuming responsibility in one of the three major airports in Greece. She considers the sector as a research vocation and is enthusiastic for aviation and more spontaneous in her reactions
Vasileios Pyrros	Male air traffic controller in his third airport position, working for 6 years in Athens Airport tower
Omiros Chios	Female air traffic controller (and married to an air traffic controller), in one of the top 4 airports in the Greek island periphery
Filippos Kozanis	Male air traffic controller with responsibility on collecting the anonymous incident/accident reports from Greek airspace and inserting them in the European Incident Database. This is the process that started on 2005 to consolidate reporting at a national level (Permanent Committee for Safety Reporting). Anonymous reports update the database at a European level and encourage the flow of reports similar to the NASA/ASRS system
Aristotelis Kastorias	A very experienced female air traffic controller who is heading the service in a major island airport. She has experienced the pressing operation of controlling air space during the occurrence of major physical phenomena like earthquakes. She has also a certified instructor on safety courses for air traffic controllers in SPOA
Alexandros Onassis	A Lieutenant General in HAF flight engineering, involved with aerospace navigation system and an expertise in satellite monitoring
Dimitrios Vikelas	An experienced male air traffic controller responsible for the localization of ICAO manuals in the Greek Flight School (SPOA). He is a member of the social media administration for a community of practice of aviators involved in language localization standards
Vassilis Konstantakopoulos	A male air traffic controller with the strongest theoretical background for those who respond in this study. He holds a PhD in aviation safety. He is an active controller in a major island airport and participates in EASA task forces
Neos Aghialos	Fighter pilot in F-16 with experience in Renegade interceptions
Palios Ellinikos	A commercial pilot who introduced the language standardization scenario after a debriefing discussion in SPOA School
Stan Dardman	A civil aviation pilot introducing the directness in language scenario, after discussing the IATA phraseology survey

Index

T. Katerinakis, *The Social Construction of Knowledge in Mission-Critical Environments*, Innovation, Technology, and Knowledge Management,
https://doi.org/10.1007/978-3-319-91014-7

The manufacturer's authorised representative in the EU is Springer Nature Customer Service Centre GmbH, Europaplatz 3, 69115 Heidelberg, Germany. If you have any concerns regarding our products, please contact ProductSafety@springernature.com

Printed and bound by CPI Group (UK) Ltd, Croydon, CR0 4YY
30/09/2026
02225573-0005